CAMBRIDGE LIBRARY COLLECTION

Books of enduring scholarly value

Earth Sciences

In the nineteenth century, geology emerged as a distinct academic discipline. It pointed the way towards the theory of evolution, as scientists including Gideon Mantell, Adam Sedgwick, Charles Lyell and Roderick Murchison began to use the evidence of minerals, rock formations and fossils to demonstrate that the earth was older by millions of years than the conventional, Bible-based wisdom had supposed. They argued convincingly that the climate, flora and fauna of the distant past could be deduced from geological evidence. Volcanic activity, the formation of mountains, and the action of glaciers and rivers, tides and ocean currents also became better understood. This series includes landmark publications by pioneers of the modern earth sciences, who advanced the scientific understanding of our planet and the processes by which it is constantly re-shaped.

The Mineralogy of Derbyshire

Born in Derby, John Mawe (1766–1829) established a successful mineral-dealing business in London and became a significant figure in the development of British commercial mineralogy in the early nineteenth century. He travelled widely, advising on mineral exploration, and gathering specimens for clients such as Charles IV of Spain. This illustrated 1802 work gives an overview of the geological features and strata of Mawe's home county. He discusses the various mineral deposits to be found in Derbyshire, and describes some of the county's mines. Drawing on observations made on his travels, he gives descriptions of important mines in northern England, Scotland and Wales, alongside remarks on geological features of interest. The book closes with a glossary of terms used by miners in Derbyshire. Mawe's well-received *Travels in the Interior of Brazil* (revised edition, 1821) is also reissued in the Cambridge Library Collection.

Cambridge University Press has long been a pioneer in the reissuing of out-of-print titles from its own backlist, producing digital reprints of books that are still sought after by scholars and students but could not be reprinted economically using traditional technology. The Cambridge Library Collection extends this activity to a wider range of books which are still of importance to researchers and professionals, either for the source material they contain, or as landmarks in the history of their academic discipline.

Drawing from the world-renowned collections in the Cambridge University Library and other partner libraries, and guided by the advice of experts in each subject area, Cambridge University Press is using state-of-the-art scanning machines in its own Printing House to capture the content of each book selected for inclusion. The files are processed to give a consistently clear, crisp image, and the books finished to the high quality standard for which the Press is recognised around the world. The latest print-on-demand technology ensures that the books will remain available indefinitely, and that orders for single or multiple copies can quickly be supplied.

The Cambridge Library Collection brings back to life books of enduring scholarly value (including out-of-copyright works originally issued by other publishers) across a wide range of disciplines in the humanities and social sciences and in science and technology.

The Mineralogy of Derbyshire

With a Description of the Most Interesting Mines in the North of England, in Scotland, and in Wales

John Mawe

CAMBRIDGE
UNIVERSITY PRESS

University Printing House, Cambridge, CB2 8BS, United Kingdom

Cambridge University Press is part of the University of Cambridge.
It furthers the University's mission by disseminating knowledge in the pursuit of education, learning and research at the highest international levels of excellence.

www.cambridge.org
Information on this title: www.cambridge.org/9781108076180

This edition first published 1802
This digitally printed version 2015

ISBN 978-1-108-07618-0 Paperback

To face title

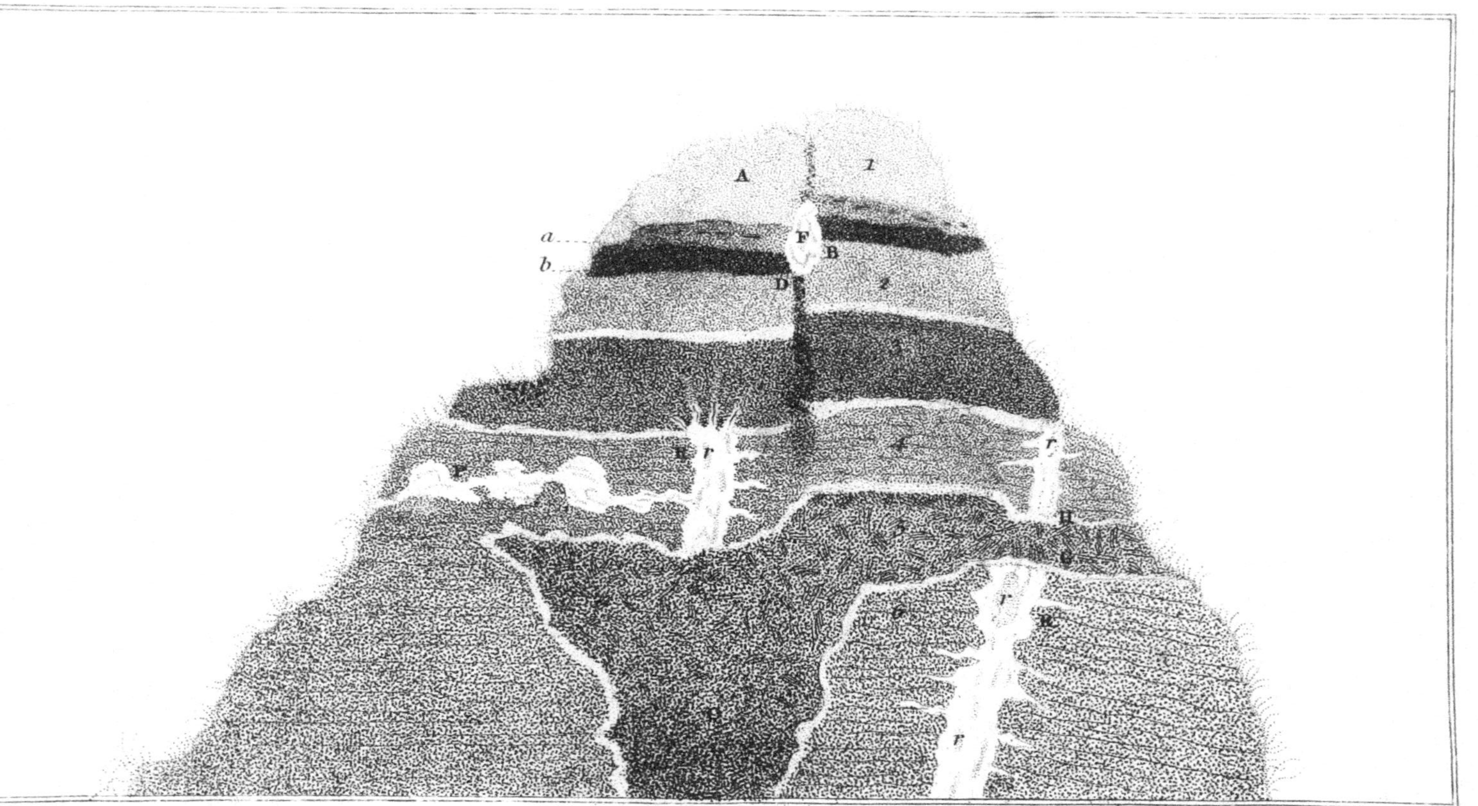

Published Jany. 1. 1802, by J. Mawe, N°. 5, Tavistock Street, Covent Garden, London.

THE MINERALOGY OF *DERBYSHIRE*:

WITH

A DESCRIPTION OF THE MOST INTERESTING

MINES

IN THE

North of England, in Scotland, and in Wales;

AND AN ANALYSIS OF MR. WILLIAMS'S WORK,

Intitled "The Mineral Kingdom."

SUBJOINED IS

A Glossary of the Terms and Phrases used by

MINERS IN DERBYSHIRE.

By JOHN MAWE.

LONDON:

Printed and Sold by William Phillips, George Yard, Lombard-street; Sold also by J. White Fleet-street, G. & W. Munn, New Bond-street, and by John Drury, in Derby.

1802.

PREFACE.

DERBYSHIRE has ever been considered as one of the chief mining counties in the kingdom, and was known to produce lead ore at a very early period. Since the Roman invasion, its mines have supplied the greatest part of Europe with their produce. It appears that the Saxons introduced their method of working the

 mines,

mines, the riches of which recompenced their labour; and the prefent mineral laws, cuftoms, and technical phrafes are derived from them. Perhaps no country yet known produces fo many veins as the mining tract of Derbyfhire; and the number of mines that have been funk in various parts is incredible. Being a native of the county, and having refided feveral years in the moft interefting part, I was applied to by a Spanifh gentleman to make furveys of the principal mines, to collect their various productions, and more particularly, fpecimens from each ftratum, defcribing their thicknefs, fituation, and pofition; in order to fhew an exact reprefentation of the mines, for the cabinet of his moft Catholic Majefty at Madrid. To afcertain

tain a correct ſtatement of the geology and mines of Derbyſhire, is a work worthy of the patronage of a Prince who enjoys ſo great a ſhare of the precious metals produced in South America: it may prove an example that may merit the attention of other potentates; for collections thus formed, diſplaying the ſtrata and their products, may lead to a more minute inveſtigation, where ſuch ſtrata may occur; the beſt means of forming opinions being by compariſon, if ſimilar mountains and ſtrata are met with, it would be very natural to expect ſimilar ſubſtances; theſe circumſtances have unfortunately hitherto attracted little notice. The ancient method of dreſſing and ſmelting lead ore is here ſtill continued, and though new veins are frequently cut, no analysis is made of their produce; it is much to be wished sufficient encouragement was given in this science, to render it worth the attention of a person of abilities to analyze mineral substances, in order to convey mineralogical information to that part of the community that is so much interested in them. For such a purpose Castleton seems to be the best situation, where such a variety of strata, mines, and mineral productions occur as perhaps no situation in this kingdom can boast. The various mines and veins of ore are of the first consequence, while the mountains around present a variety of strata worthy the attention of the geologist.

Freyberg and Schemnitz, the present theatres

theatres of mineralogical knowledge had beginnings; is it not to be regretted that no inſtitution for ſuch information is eſtabliſhed in this kingdom, the riches of whoſe mines have ſo long been celebrated?

Students attend the mineralogical lectures at Freyberg and Schemnitz from all parts of the world, and they are as much famed for the ſtudy of mineralogy, as Rome was for the fine arts.

Having frequently viſited moſt of the mines in this kingdom, I have been repeatedly ſolicited to publiſh the obſervations I have made, with a view to guide the traveller to the moſt intereſting points, and to deſcribe thoſe objects to the mineralogiſt as they are preſented by nature; as an obſerver addicted to no theory, I leave the ſcientific to form opinions agreeable to their own ſentiments. I now beg leave to ſummit this eſſay towards a deſcription of the mines in Derbyſhire, &c. to the public inſpection: conſcious I am that the plainneſs of the language may not be well ſuited to the literary world, but I hope the candid reader will excuſe it, truſting it is the beſt adapted to explain the ſubject on which I have treated, and fully acknowledging my want of abilities, as an author unaccuſtomed to compoſition. I am afraid it is impoſſible to avoid tautology in giving a deſcription of mines and their concomitant circumſtances, and in my endeavours to render them more eaſily underſtood, I may probably have had recourſe to ſome degree of repetition; if ſo, it has been in order to explain my ideas

with

with more preciſion, my ſole view being to induce others to inveſtigate this county more minutely.

It was my intention to have given a deſcription of the mines in Cornwall and the weſt of England, and their products; but being engaged in other purſuits, I muſt defer it until a more favourable opportunity.

CONTENTS.

ERRATA.

Page		line		for	read
Page	11,	line,	15,	for ſhiſtus	read ſhiſtoſe
— —		—	18,	— prites	— pyrites
—	51,	—	12,	— calcedoney	— calcedony
—	58,	—	6,	— Fagus	— Faujas: *and omit the words*, "Toad-ſtone containing lead ore.
—	84,	—	9,	— Scarſdales	— Scarſdale
—	86,	—	2,	— luminated	— laminated
—	95,	—	6,	— pirites	— pyrites
—	96,	—	19,	— amorphus	— amorphous
—	111,	—	20,	— where	— were
—	134,	—	8,	— ſchiſtus	— ſchiſtoſe
—	137,	—	14,	— of a,	— of
—	—	—	18,	— dot	— not
—	141,	—	*note*	— Glenco	— Glencro
—	142,	—	*laſt*	— is is	— is
—	148,	—	15,	— was	— were
—	161,	—	13,	— Eaſt	— Weſt
—	167,	—	&c.	— Paris	— Parrys
—	—	—	10, 11,	— quartoze	— quartzoſe
—	181,	—	17,	— rock among	— rock: among
—	183,	—	19,	— Fourth	— Forth
—	186,	—	8,	— Lead	— Leod
—	186	—	12,	— ſupporting	— ſuppoſing
—	187,	—	12,	— Daven Jaur	— Davenfawr
—	189,	—	11,	— vein or ſtreek	— veins or ſtreeks
—	—	—	15,	— coal	— cone
—	194,	—	3,	— granatic	— granitic
—	195,	—	12,	— brecica	— breccia

CONTENTS.

SECTION I.

SECTION II.

SECTION

SECTION III.

SECTION IV.

SECTION V.

SECTION VI.

SECTION VII.

SECTION VIII.

SECTION IX.

SECTION X.

SECTION XI.

SECTION XII.

SECTION XIII.

SECTION XIV.

SECTION XV.

SECTION XVI.

SECTION XVII.

SECTION XVIII.

To face p. 1.

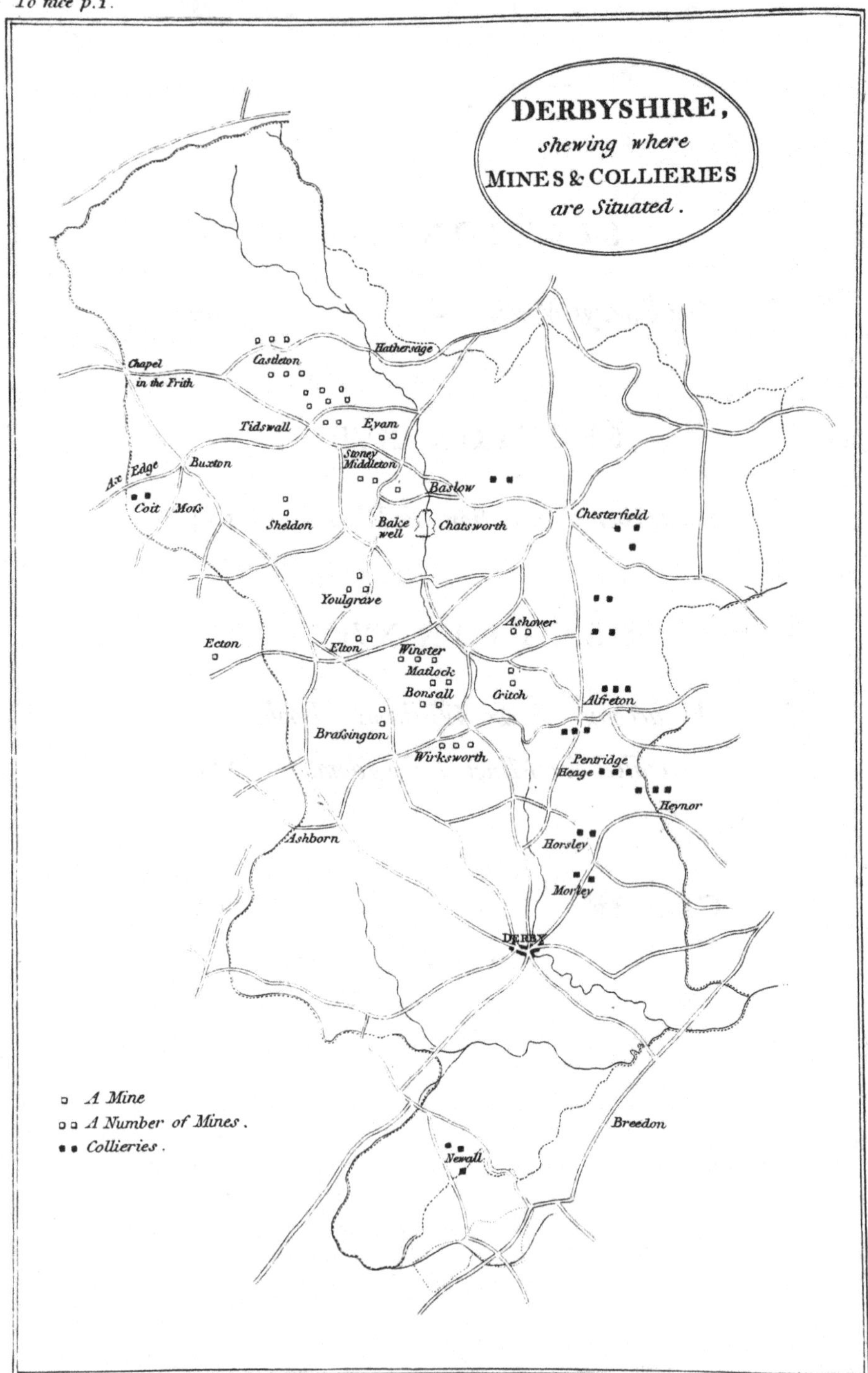

Mutlow Sc. Rufsell Co.

SECTION I.

Curiosities of Derbyshire; particularly near Castleton.

ON approaching Derbyshire from the south, the eye of the traveller, fatigued with level uniformity, is agreeably relieved with the prospect of mountains. For here begins the chain which has been called the English Apennines; and which, forming, as it were, a root in Derbyshire and part of Cheshire, afterwards sends forth a trunk, which running

ning due north, branches into the mountains of Weſtmoreland, Cumberland, and Northumberland.

Theſe mountains have been ably deſcribed by Dr. Aikin, Mr. Houſman, and others; and theſe remarks ſhall be confined to their mineralogical productions; after a few general obſervations on ſome parts of Derbyſhire, and in particular on the vicinity of Caſtleton.

Derby is ſituated in the ſouthern part of the county, while the chief mountains, and Caſtleton, are in the north. The capital of the ſhire is a well built town, and of late has received conſiderable improvements. It is ſituated on the river Derwent, over which there is a new ſtone bridge. There are five parochial churches, of which All Saints, the principal, ſtands in the centre of the town; a handſome modern edifice, the roof being ſupported by elegant columns of the Doric order,

order, and of confiderable fize. But the tower is ancient Gothic, richly ornamented, and about 180 feet high. Here is the ancient burial place of the Devonfhire family, and there are fome good monuments of the houfe of Befborough. Derby has communications with many canals, and navigable rivers, and is founded on a ftratum of grit ftone; beds of gravel, compofed of filiceous rounded pebbles, of various fizes, are frequently incumbent on it in the neighbourhood.

The filk mills, erected by Sir Thomas Lombe, are fine buildings of confiderable extent, and giving employment to numbers of men, women, and children. The proprietor is always ready to impart information to the curious vifitor. The firft mill that was built for Sir Thomas is now converted into a manufactory for fawing, turning, and polifhing the fluor fpars; the whole of the operations being conducted by machinery, ſubſervient to the power of water.

The porcelain manufactory, belonging to Meſſrs. Dueſbury and Kean, is worthy of the patronage of the illuſtrious family who have honoured it with their approbation. Here the whole proceſs of making what we call China may be ſeen; and the beautiful painting and gilding have conferred on this manufacture a great reputation.

There are alſo many cotton mills, the principal belonging to Meſſrs. Strutts: and a rowling and ſlitting mill, where iron plate is tinned; with a manufactory of white lead, and one of red lead at Darley near Derby.

On the road to Matlock, four miles from Derby, is the magnificent ſeat of Lord Scarſdale, called Keddleſtone Hall, with a park, wood, and gardens, which are deſervedly admired. Matlock, a bathing place, is celebrated for its romantic ſituation. Dovedale, near Aſhborne, is a beautiful valley, through which runs the rapid river Dove, among rocks and woods, uncommonly ſtriking and picturesque.

Buxton is well known by its hot baths, and the beautiful creſcent, built for the public accommodation by the duke of Devonſhire.* This place is much frequented; and throughout the whole county the traveller may depend on good roads and excellent inns.

Chatſworth was once eſteemed *among* the wonders of Derbyſhire, being a ſummer reſidence of the duke of Devonſhire; is very ſtately and ſpacious, with delightful gardens, pleaſure grounds, and water works.

Other curioſities of Derbyſhire are the grand cavern called Peak's Hole, the Elden

* The traveller will be much ſurpriſed to ſee a building in this remote part of the kingdom that rivals the beauty of Palmyra.

Hole in the Peak foreſt, and the ebbing and flowing well near Caſtleton. Monſaldale, near Aſhford, is a beautiful ſmall valley, where nature ſeems to have exerted herſelf, to contraſt and diverſify the ſcenery, ſo as to equal any thing of the kind in the kingdom.

Proceeding north to Caſtleton, the moſt ſtriking object is the caſtle, which by the Romans was called *arx diaboli*; it ſtands on a rock of limeſtone, inacceſſible in every direction, except to the ſouth. The buildings encloſe an area of larger extent than would be expected; and from the foot of the hill extends on each ſide a ditch which ſurrounds part of the town, being three yards wide and two deep. Heads of arrows are frequently found; and alſo Roman coins. I have in my poſſeſſion a Roman celt of braſs found here, about five inches in length, weighing about a pound. It is evident that the Romans worked the lead mines

mines here, as a bar of lead was found marked with the name of one of the emperors; and which I believe is now in the museum of Mr. Green at Litchfield. Near Castleton are many fine springs of water; and in the neighbourhood of Bradwell is a warm salt spring, which has not yet been analysed.

About five miles from Castleton, and on the road to *Chapel en le Frith*, is the ebbing and flowing well, at the bottom of a limestone hill, and several yards in circumference. After it ebbs there is scarce any water, except at the sides which first begin to flow. In wet weather it flows and ebbs several times in an hour: while it flows the water boils up with great violence, in a number of places, for five or six minutes; then it ceases, the water runs off, and after about ten minutes it begins to flow again. In dry weather it does not flow so often.

Proceeding a mile further, and at the town end of *Chapel en le Frith*, is a new piece of mechanifm, called the inclined plane, a name which explains its nature. It is formed on the fide of a mountain, in order to convey limeftone to the Manchefter canal. The carts hold about three tons each, and their velocity is regulated by mechanical principles. While the loaded carts defcend, the empty ones afcend to be filled again. This limeftone forms a confiderable article of commerce, being tranfported many miles, and efteemed of a very fuperior quality. The noted cavern of Peak's hole has been fo often defcribed that any further account would be fuperfluous: but a fhort defcription may be allowed of another wonder of the Peak, not fo generally known, concerning which marvellous ftories have been told, and this plain account may at leaft fave the reader from impofition. Elden

hole,

hole in Peak foreſt, is a chaſm or fiſſure on the ſide of a limeſtone mountain, about 30 yards in length, and from 7 to 9 yards wide. The form is irregular, the depth about 60 yards, the ſtratum ſeparating at the bottom, with ſome communications of inconſiderable extent. Any miner could go down with eaſe, for a ſmall compenſation; he would call it a *ſhake*, *ſwallow*, or *opening*, as ſhall afterwards be explained.

SECTION

SECTION II.

Account of the Strata in Derbyſhire.

HAVING thus given a curſory idea of ſome intereſting objects on the ſurface of Derbyſhire, let me next be permitted to accompany the reader under ground, and to explain the general formation of the ſtrata, ſuppoſing a mountain to be vertically divided. This appearance will beſt be underſtood by referring to the annexed plate, which I ſhall proceed to illuſtrate, after obſerving that the ſtrata in Derbyſhire are ſingularly curious, and perhaps unlike any thing to be found on the continent, being conſidered by foreign mineralogiſts as often preſenting exceptions from the general rules obſervable in continental mines.

No.

No. 1. reprefents the fummit of a hill. A is argillaceous grit: *a* fhews irregular beds of argillaceous and fulphureous iron ores attendant on coal: *b* is coal lying in laminæ under the argillaceous grit. The depth of thefe ftrata is as follows.

Although argillaceous grit is generally above coal in this county, it is not to be underftood that it is invariable fo; for a variety of fubftances which frequently appear in great confufion fometimes are fuperincumbent: as vegetable earth, gravel or rubble, compofed of quartzofe pebbles, clay, and pieces of argillaceous fandftone. Indurated clay; a fpecies of fhiftus fand ftone in laminæ, blue clay; femi-indurated black earth or fmut; argillaceous iron ore; and thin beds of prites and fhiftus. Thefe fubftances have frequently various names, as metal, bind, ratchel, clunch, &c. They at all times indicate coal; and though coal is

found

found under, and in a variety of ſtrata in other parts of Europe, yet in this county it has been hitherto confined to the argillaceous.

No. 2. Siliceous grit, forming a ſtratum of unequal thickneſs, ſometimes exceeding 120 yards.

No. 3. Shale or ſhiſtus, appearing like an indurated clay, of various thickneſs, ſometimes equal to No. 2.

No. 4. Limeſtone of various thickneſs, from four fathoms to more than 200, and not cut through.

No. 5. Toadſtone, which frequently divides the limeſtone.

No. 6. Limeſtone, beyond which no mine in Derbyſhire has penetrated.

Each ſtratum is ſeparated by a ſmall ſeam of clay, or marl, differing in thickneſs from two or three inches to two feet; and of various colours, from the ochre yellow to the

the brown, and aſh green. It is worthy of notice that whatever ſtratum appears the uppermoſt, this repreſentation will ſhew the ſubſequent arrangement; a circumſtance deſerving attention in mining countries: for by the knowledge of the upper ſtratum the ſkilful miner is enabled to form an idea of what may be found underneath, whether coal, iron, &c. &c.

The ſurface of the valley of Caſtleton is rubble, compoſed of broken fragments of various ſubſtances, ſome as ſmall as coarſe gravel, reaching to the depth of a few fathoms, as repreſented in the plate.

I ſhall now proceed to examine the ſubſtances that compoſe each ſtratum, and thus endeavour to point out the uſe of mineralogical knowledge, as many gentlemen for want of receiving ſome information on that intereſting ſcience, have been expoſed to the arts of their agents, and have ſuffered

great

great impoſitions and loſs. But now that mineralogy is becoming a faſhionable ſtudy, we may expect to ſee great improvements in this important branch of natural hiſtory.

Let us now return to a more mintue conſideration of the ſtrata above delineated.*

No: 1. Argillaceous grit forms the uppermoſt ſtratum, and is more or leſs thick, as the ſurface is more or leſs uneven. It is an aſſemblage of ſand, and adventitious matter, in a baſe of argil; *fracture* granular; of a dull colour; ſmell earthy when breathed on: does not efferveſce with acids; does not take a poliſh; may be eaſily ſcraped with a knife; has often browniſh red veins; and is ſometimes ferruginous, which renders it heavier. By expoſure to the atmoſphere it decompoſes.

* Tablets, compoſed of the ſubſtances themſelves, in their natural order, may be had of the author, forming a portable and intereſting picture of the geology of Derbyſhire.

This

This ſtratum generally indicates iron ore, which is frequently found under it in laminæ and nodules. The argillaceous iron ore is the moſt common: *a* repreſents a thin bed of it, of a brown colour, and compact nature; ſmell earthy; yields about 30 per cent. ſeldom more. Nodules of this ore are frequently found, which eaſily divide, and ſhew very fine impreſſions of plants, flowers, coralloids, and ſhells. The ſtrata of argillaceous grit and iron are generally incumbent on coal, as at *b*, which repreſents coal lying in laminæ, unequal in quality and thickneſs. It frequently abounds with pyrites or ſulphuret of iron, and argillaceous iron ore in nodules: fracture generally ſplintery, laminated, ſometimes regular, with a bright gloſs, and very brittle: contains much ſulphur and petroleum.

Coal is found at Newhall, about ten miles ſouth of Derby; it is there covered

with

with a variety of earthy ſubſtances, the ſtrata being of various thickneſs, in different ſituations where the mine is ſunk. Firſt vegetable earth a few inches, then 12 feet argillaceous blueiſh earthy matter, 44 to 50 feet decompoſed black earthy ſhiſtus, a bed of 6 feet of ſhiſtoſe hard coal, under which is a ſtratum of argillaceous indurated clay, from 10 to 12 feet, which is incumbent on a bed of fine coal, 8 to 10 feet thick.

In the neighbourhood, to the north eaſt, is a large mountain of limeſtone, containing a conſiderable portion of magneſian earth, at Breedon, on the edge of Leiceſtershire, uſed for land and building purpoſes; its colour being rediſh grey: in it are ſparry veins, and ſometimes ſmall ſtrings of galena. Proceeding north, the coal does not make its appearance until you arrive north eaſt of Derby, a diſtance of twelve to fourteen miles; here a large tract of country is

enriched

enriched by this valuable commodity, as at Morley, Hallam, Smalley, Denby, Heynor, Pentridge, Alfreton, Chefterfield, Baflow, and many other places, amongft which are fome iron works.

The coal is found at various depths; and where a horizontal gallery can be driven into the coal, it is certainly much more convenient and lefs expenfive than the general mode of finking fhafts.

The national benefit arifing from this article is beyond eftimation; canals are cut to tranfport it into thofe diftricts in which no coal is found, by which many thoufands find employment. Coals may be bought at the mine for 5s. 6d. per ton, or at 10s. per ton conveyed a few miles.

The great improvement which the iron manufactories of this country have received by charring or coaking the coal, now frequently adopted, gives reafon to hope that they

they will ſoon rival thoſe of Sweden and Ruſſia. The Engliſh iron, twenty years ago, ſcarcely deſerved the name, as it could not be worked into any article of fineneſs; but ſuch is the improvement, that we now have but ſmall demand for foreign iron.

It is not an eaſy matter to determine the extent of this improvement, as iron works are ſo conſideraby increaſing all over the kingdom; and at ſome diſtant period we probably may poſſeſs our mines of coal when the foreſts of the northern powers may perhaps be conſumed: ſuch is the poſſibility of affairs; nor is it extremely improbable but this country may at ſome future period export her iron to the nations that half a century ago excluſively ſupplied us.

Our iron bridges are a ſpecies of architecture of which this empire alone can boaſt. Iron in its various ſtates is ſo applicable to the uſe of man, that it would be

daily

difficult to form limits to its application; and I am credibly informed that the demand daily increaſes. We have a few works in this kingdom in which charcoal is uſed in the making of iron; the iron thus produced is equal to the beſt Swediſh, and probably we ſoon ſhall procure from it as fine ſteel.

Coal frequently emits while burning a liquid bituminous matter; and ſhiſtus is frequently ſo penetrated with aſphalt as to burn until the inflammable matter is volatalized. In this country pieces of coal may be got very large, weighing more than three or four hundred pounds. Veins of ſulphate of iron frequently occur; and in two or three inſtances lead ore has been found in it. When the ſulphate of iron has appeared in abundance, and the ſituation convenient, copperas works have been eſtabliſhed.

The culm, or ſmall coal, is in many caſes of no value, and may be taken away from the mine gratis.

Coal prefents feveral varieties, and is more or lefs interefting; fome are fhiftofe, hard and fulphureous; others are foft, and quickly confume.

The beft coal is generally of the leaft fpecific gravity, and of the brighteft black colour, finely laminated, and on burning leaves the leaft afhes.

The cannel or candle coal is very compact; fracture fplintery; it is lighter than the other variety, and is fonorous when ftruck; frequently explodes when heated, and burns with a luminous flame; its colour is jet black; it is capable of a fine polifh; it feems to contain more carbon and lefs fulphur; it is fometimes found under and in connection with the common varieties.

Coal is fometimes in contact with afphalt and indurated bitumen.

Coal mines are of various depths; and coal often baffets * out to the grafs. The

* See the gloffary at the end.

ftratum

ſtratum is frequently broken, when the workmen meet with a fault, *mear*, or *lum*, which is a cavern filled up with clay, or rubble, diſlocating the ſtratum of coal. In ſuch caſe the coal is ſometimes lifted up, ten or twenty yards; or as much thrown down. See the coal ſtratum at B, (ſee plate 1.) and the fiſſure at F where the coal is thrown down at D. In ſearching for coal, ſtreams of water, after heavy rains, ſhould be examined; and when it is found, the moſt eaſy method of working ſhould be adopted, that an article of ſuch general uſe may be rendered as cheap as poſſible. It is the grand ſource and root of all our manufactures, and of the firſt national conſequence; whence too much encouragement cannot be given to ſearch for this neceſſary article, where it has not yet been found. It would prove an inexhauſtible ſource of wealth in the Highlands of Scotland: agriculture would flourifh, the arts and manufactures would be extended; and from its appearance might be dated the riches of that country.

Coal is frequently found under a variety of fubftances, commonly appearing in the form of ftrata, and called by the colliers *under-foil*, *gravel*, *bind*, *clutch*, *hardftone*, *metal*, *plate*, &c. as has been before mentioned, which are fometimes only a few inches thick, in others feveral feet; but generally the grit is fuperincumbent.

No. 2. (fee plate 1.) The ftratum of coarfe filiceous grit, extending at the moft about 120 yards, and variable in its appearance and texture. It forms the uppermoft ftratum in Wirkfworth Moor, Cromford Moor near Winfter, the Eaft Moor, Birchover, Matlock town, the Edge fide Hills, from Eam to Caftleton, of Mam Tor, and in many other places.

It is an affemblage of coarfe quartzy pebbles

pebbles of various ſizes, ſeldom exceeding a quarter of an inch diameter; ſome are in part cryſtalized with ſharp angles, others are rounded; it is very friable near the ſurface, and ſometimes contains adventitious matter: it is not ſtratified. It gives fire with ſteel; reſiſts acids; and is often coloured by iron: fracture irregular, and does not take a poliſh. In this ſtratum are varieties of grit ſtone in laminæ; ſome are called freeſtone, and uſed in buildings; others called millſtone grit, and uſed for millſtones. A particular variety is laminated with mica, being an excellent ſubſtitute for ſlate, and uſed in forming the roofs of buildings, whence it is an article of commerce. This variety is ſomewhat elaſtic, and eaſily divides with a knife. Frequently it contains cryſtallized fluor, and barytes, and is incumbent on ſhiſtus or ſhale, from which it is ſeparated by a thin ſeam of clay.

In general it is rare to find veins in this ſecond ſtratum, but there are ſome inſtances of lead ore being found in it.

No. 3. The ſtratum of ſhale or ſhiſtus is not ſtratified; it ſometimes is 120 yards in thickneſs; and is the uppermoſt ſtratum in many of the valleys, where the mountains on one ſide are grit, and on the other limeſtone: the ſhale betwixt the two entering the edge of the limeſtone, and paſſing under the grit. The hot waters of Buxton are found in this ſubſtance.

It is of a dark brown or blackiſh colour, bituminous, and appears much like an indurated clay: it does not contain vegetable impreſſions, though ſometimes impreſſions of marine ſubſtances are found in it much impregnated with pyrites. It is not genenerally conſidered as a ſtratum fertile in veins of lead ore, though ſometimes that ſubſtance is found in it, for being incumbent

bent on limeſtone, the veins ſtrike from it into the ſhale, and carry lead ore with them to ſome diſtance. In its ſparry veins are frequently cavities, called *lochs* by the miners, which are incruſted with fine and rare cryſtallizations of calcareous ſpar in great variety.

By expoſure to the atmoſphere, this ſhale decompoſes in *laminæ*: its fracture is dull: it abſorbs moiſture: contains ſulphur burning with a blue flame, and becoming of a rediſh brown colour: frequently reſiſts acids, but ſometimes efferveſces ſlowly: contains nodules and thin beds of pyrites. The waters paſſing through it are chalybeate, and frequently warm. It is incumbent on limeſtone, and is ſeparated from it by a thin bed of clay. When it approaches the limeſtone, it of courſe efferveſces with acids; in ſome caſes even containing a

large

large portion of calcareous earth; the lime-ſtone in return partaking of its dark colour ſeveral feet from where they are in contact.

SECTION

SECTION III.

The subject continued. Further account of the Strata of Derbyshire, particularly of the Limestone and Toadstone.

HAVING thus discussed the superior strata, I shall next proceed to others which are more interesting to the miner and geologist.

No. 4. (see plate 1.) The first stratum of limestone is regularly stratified, and varies considerably in depth, being in some places thin, while in others, as already mentioned, it is extended to more than two hundred fathom. It forms the uppermost stratum east of Wirksworth, at Matlock, at Winster, Ashford, Eyam, Buxton hills, Monyash, and south of Castleton. It lies in laminæ, more

or

or lefs thick, and is frequently feparated at irregular diftances of feven or five fathom, &c. by a marl containing adventitious fubftances, and in fome places only a few inches thick, while in others the marl is two feet.* The whole of this ftratum is compofed of marine exuviæ, which fhew it not to be what is underftood by primitive limeftone. On the furface of this ftratum is fometimes found rotten ftone, particularly near Wardlow Mire and Afhford, which appears a decompofed argillaceous fubftance containing oxyd of iron. It feels fmooth, and is much ufed for polifhing brafs in the manufactories at Sheffield.

This ftratum abounds with a variety of fhells, entrochi, coralloids, madrepores, &c. The vallies often contain ratchell or rubble,

* If more notice were taken of the divifions in the limeftone, by the marl and adventitious matter interpofing, probably the formation of calcareous earth might be better explained.

a cona confufed mafs of various fubftances, of different fizes, collected by their rolling from the mountains at various periods. Sometimes the hills on one fide are limeftone, and on the other grit or fhiftus. For inftance, the limeftone, which to the fouth weft of Caftleton is uppermoft, is 300 yards below, under the oppofite mountains, which are of grit, incumbent on fhale. As neither the ftrata of fhale nor grit make their appearance on that of limeftone, to the fouth weft of Caftleton, nor in many other parts of this country, it was the opinion of Mr. Whitehurft, that fuch ftrata were diflocated and thrown into confufion.

The limeftone forms a variety of beautiful marbles; near Wirkfworth fome are found in thin ftrata, of a light ftone colour, full of marine remains, and ufed for paving, flooring, &c. called Hopton ftone. Near Monyafh, and at Foolow, a beautiful variety is found

found in a confiderable quantity, of a cheerful colour, inclining to the brown red, and full of large marine figures in all directions, which on being cut appear white, and afford a pleafing contraft. This beautiful marble is ufed for chimney pieces, and other ornaments. Near Wetton, a variety is found, of a darker colour, and prefenting very fmall figures, whence it is called bird's eye marble.

In various parts black marble is found in laminæ, being coloured by iron and petroleum, which is frequently found to pervade the mafs. It burns to a white lime, which forms a ftong cement. All the varieties are fœtid when rubbed with a harder fubftance. The coralloids that are found in the black marble have a very pretty ftarry or ftellated appearance, but fuch pieces are not common.

A filiceous fubftance called chert, ufed by

by the potters, is found in the limeſtone ſtratum, in large detached maſſes and thin ſtrata, near Caſtleton, at Dirtlow, at Bradwell, in Middleton dale, in Peak foreſt, Matlock, and various other places. This ſubſtance is full of marine figures, and animal remains; in which reſpect it reſembles the limeſtone, as though it had undergone a tranſition into petroſilex, or what the French call keralite. I have ſpecimens, partly ſiliceous, and partly calcareous. The ſhells in this ſubſtance, and in the limeſtone, are full of calcareous cryſtallizations, and ſometimes contain bitumen.

In this large calcareous ſtratum are many caverns, particularly that wonderful work of nature, Peak's Hole.

The limeſtone in the Peak foreſt is the beſt: the fracture ſcaly bright; it is compact; and ſonorous when ſtruck. It burns to a fine white lime, loſing about thirty per cent.

of

of the carbonic gas during the operation, which occupies about thirty hours of a ſtrong fire. It is burnt in conical kilns of various ſizes. Irregular maſſes of lime-ſtone, conſiſting of fragments cemented together by infiltrated water, are ſometimes found, with cryſtallized calcareous ſpar, &c. in the interſtices.

This ſtratum is the moſt intereſting to the mineralogiſt, for in it are found the principal veins, containing galena, ſulphuret and native oxyd of zinc, a variety of ochres, fluors, barytes, calcareous cryſtallizations, pyrites, &c.*

I may here be permitted to give a ſhort account of the metallic veins, as they occur in Derbyſhire. They are chiefly divided into two varieties, *rake* or perpendicular

* The great copper mine at Ecton is in this ſtratum; and in other parts of England I have ſeen copper ore and iron ores, in conſiderable quantity, in the limeſtone.

veins,

veins, as at R; and *pipe* or flat veins as at P. The rake veins are in different directions. Near Castleton they generally run from east to west, and are traced, or discovered, from the surface. They are not exactly perpendicular; but *hade*, or incline, about one foot in ten, sometimes to the north, and sometimes to the south. There are veins that have a more northerly or southerly direction, and are then called *cross veins*. Sometimes they intersect each other; and where they unite they are generally very rich. Small veins, usually called *strings* or *scrins*, often extend from the rake, and take various directions. All are worked as long as they are found profitable: and the intermediate substances that divide them are called *riders*. (see *r*.)

The rakes generally form a strait line, and very rarely assume a bent direction. When separated, which is sometimes the

 case,

caſe, by *clay*, *bind*, or *toadſtone*, they are obſerved (upon being again diſcovered below) not to be perpendicular, but to be thrown to one ſide, according to the *hade* of the vein, and the diſtance of ſeparation, and are thence ſaid to *leap*. From this obſervation it does not appear adviſeable to ſink a ſhaft or ſump perpendicular from where the vein was cut off, in order to find it again, but to make a croſs cut for ſome fathoms that way which forms an obtuſe angle with the vein; by that mean you will be certain on ſinking, after cutting acroſs a proper diſtance, to find the vein again. Sometimes one part of a vein will hade, and another part ſtand perpendicular, as in Mandel mine, near Sheldon. The rakes differ much in ſize, in the ſame vein, and are ſubject to *twitches*.

The principal veins near Wirkſworth are called Yolk cliff rake, Hollyhole vein,

Rantor

Rantor Tor, Orchard, Ratchwood, Pen's rake, Grey Mare, Samuel rake, and many others: ſome range north and ſouth, and others eaſt and weſt; and ſometimes ſeveral veins unite, and form a very large one: beſides which there are many other veins in the neighbourhood of Matlock, Bonſal, Winſter, Elton, Youlgrave, and other places.

Near Caſtleton, the moſt northerly vein of lead ore in the county is Oden, a large work. A number of veins of leſs note, are in a mountain called the Long Cliff (and a ſtrong pipe vein) which extends to the Red Seats and Mr. Eyre's Park, Dirtlow and Pindar; and ſouth is Moſs rake, Hell rake, Shuttle rake, Hucklow and Tideſlow rake; Seedlow rake; at Wardlow is Longſon edge vein, and Bright ſide at Calver.

The pipe veins, or flat works, as at P. form another variety. They do not follow

 any

any regularity, or inclination of the ſtratum, but fill up fiſſures; having ſound rock for *roofs* and *ſoles*, the vein running more or leſs horizontally. They are ſometimes of great magnitude, twenty or thirty yards wide, and ſometimes ſo ſlender as not to exceed two inches.

Pipe veins are always attended with a conſiderable portion of clay, which, when the vein becomes imperceptible, will be a ſure guide to follow; and from the appearance of a few inches of clay only, purſuing it a few feet, the vein has been found of conſiderable extent. Such is the irregularity of pipe veins. The gangart of the pipe is different from that of the rakes, and they moſt commonly have the toadſtone in the vicinity, either above or below.

The principal pipe veins are Yate ſtoop, near Winſter; Hubberdale, near Money-

aſh;

aſh ;. Watergrove, Millermine, and Lane-head at Caſtleton.

In the neighbourhood of Wirkſworth, Matlock, Bonſal, Caſtleton, &c. are many veins containing blende or black jack, ſulphuret of zinc ; calamine, lapis calaminaris, or native oxyd of zinc ; barytes, calcareous ſpar, &c.

At Braſſington Moor, north weſt of Wirkſworth, are carbonates of lead, irons, ſteatite, calamine, and blende. The white lead ore is commonly in *lums*. Theſe minerals are chiefly found under looſe ſandy adventitious matter, which might deſerve the attention of the geologiſt. Black wad, an ore of manganeſe, is found near Youlgrave. The fluor ſpar mines at Caſtleton are intereſting to curioſity, as they ſhew ſuch a variety of *lums*, or broken ſtrata, filled up with adventitious matter, as are perhaps no

 where

where elſe to be met with; and produce the greateſt variety of fluors in the world.

In this limeſtone ſtratum are frequently found openings or caverns, which are commonly called *ſhakes*, or *ſwallows*. They are large fiſſures, the depth and communications of which cannot be aſcertained, but they are of great ſervice in ſeveral mines, as receptacles for the *deads*, or rubbiſh, and as aqueducts to carry off the water.

I now come to a ſtratum which has excited great attention among geologiſts and mineralogiſts, foreign and domeſtic. No. 5. repreſents *Toadſtone*, by which name various ſubſtances have been denominated, ſome having the appearance of baſalt, with equal hardneſs; while others are of various colours, full of holes and quite ſoft. When a ſubſtance is met with, intervening the limeſtone ſtratum, but different in colour and texture from the generality of limeſtone;

ſtone; it is here generally called *cat dirt*, *channel*, or *toadſtone*. How far they may prove to be what is generally underſtood by the name of toadſtone, the reader will determine.

This ſtratum is very irregular in its appearance, thickneſs, and direction. In the neighbourhood of Wormhill, Aſhover, Buxton, Caſtleton, and various other places, it appears at the ſurface, being the uppermoſt ſtratum. It is generally of a dark brown colour, with a greeniſh tinge, ſuperficially full of holes; but at a greater depth it is more compact, and the holes ſeem to have been filled with calcareous ſpar, and ſometimes with green globules. Fracture irregular; eaſily ſcraped with a knife; but this ſoft variety appears to me to be in a ſtate of decompoſition.

The harder variety is found in an irregular column, in the *cave** at Caſtleton. This is as hard as any baſalt I have ſeen; is compact; contains hornblende; and ſome patches or ſtreaks of red jaſper. This ſort is alſo found near Buxton, containing zeolite and calcedony. In no inſtance does it preſent veins of lead ore, or any mineral ſubſtance, at leaſt in Derbyſhire. The appearance of this variety aſſumes ſo many different characters according to its ſtate of decompoſition, that it is very difficult to trace its origin. The baſalts I met with at Saliſbury craggs, near Edinburgh, alſo near Glaſgow, in the iſland of Mull, and in Staffa, when in decompoſition, have every appearance and exterior characteriſtic of ſome of the varieties of the Derbyſhire amygdaloid.

It has not any appearance of ſtratification.

* This is a deep ravine at the back of the caſtle; and muſt not be confounded with the cavern at Peak's hole.

It

It resists acids. No vegetable nor marine figures have been found in it; nor any characteristic mark to evidence that it was formed at the same time, or by the same means, as the limestone stratum which it divides. The exterior, or what has been exposed to the atmosphere, resembles a scoria or vitrified mass; but this appearance may proceed from its containing oxyd of iron: and it absorbs moisture. The fracture of a dull colour; earthy smell when breathed on. It certainly contains iron in a large proportion, which is easily attracted by the magnet after torrefaction. It divides the limestone stratum, intersecting and cutting off the veins of ore as at H, which are again found by cutting through it to G. It frequently fills up fissures of great depth, as at O, while at a small distance from such fissures, it is only a few fathoms in thickness.

A de-

A defcription of a fubftance called *channel*, or *cat dirt*, or *toadftone*, containing lead ore, and mentioned by Faujas de St. Fond, Werner, Kirwan, &c. fhall be given in its proper place; and on examination it will prove to be another fubftance. Miners call every fubftance in the limeftone ftratum, differing in colour, &c. by thofe names; and travellers too frequently adopt their language, and rely on their information, without examining the fubftances themfelves.

No. 6. is another ftratum of limeftone, refembling No. 4 in every particular, which render a defcription unneceffary.

This brief account will give the reader an idea of the general produce of Derbyfhire. The miners have laws peculiar to themfelves, of ancient date, and rigidly obferved. The lead ore throughout Derbyfhire is fuppofed to be what is called potter's

ore.

ore. There is not one refining furnace. The reverberating furnace is in moſt general uſe.

The calamine and blend are got in abundance at Bonſal, Wirkſworth, Matlock, Caſtleton, &c. being bought by the braſs founders. There is a houſe where it is calcined at Cromford, whence it is conveyed to Stone, in Staffordſhire, &c. The iron works are ſmall, and not numerous, though there be ſeveral at Cheſterfield.

The mines throughout the county were formerly much richer than they are now, and produced the fineſt cryſtallizations.* Near

* It is worthy of obſervation, that the veins are poorer, in general, the deeper they are worked, which may ſerve to ſupport the opinion that veins are not formed deeper than the cruſt of the earth; but that remains ſpeculative at preſent, as does the manner in which they are filled, more eſpecially when we know ſome of them are worked under an immenſe ſtratum, that does not even bear any kind of vein. Veins, although they appear at the ſurface, yet they are ſeldom rich until they get a conſiderable depth, and where the ſtratum forms

Near the New Haven, on the road from Aſhburn to Buxton, is a vein of argillaceous carbonate of lead, called the Wheat ſtone, and alſo ſome good white clay.

ſorms a rock of the greateſt ſolidity. The almoſt continual attendant on large veins of ore, of every deſcription, is a conſiderable quantity of water, and ſcarce is any good mine worked that does not ſuffer inconvenience from it.

SECTION

Pl. II. to face p. 45.

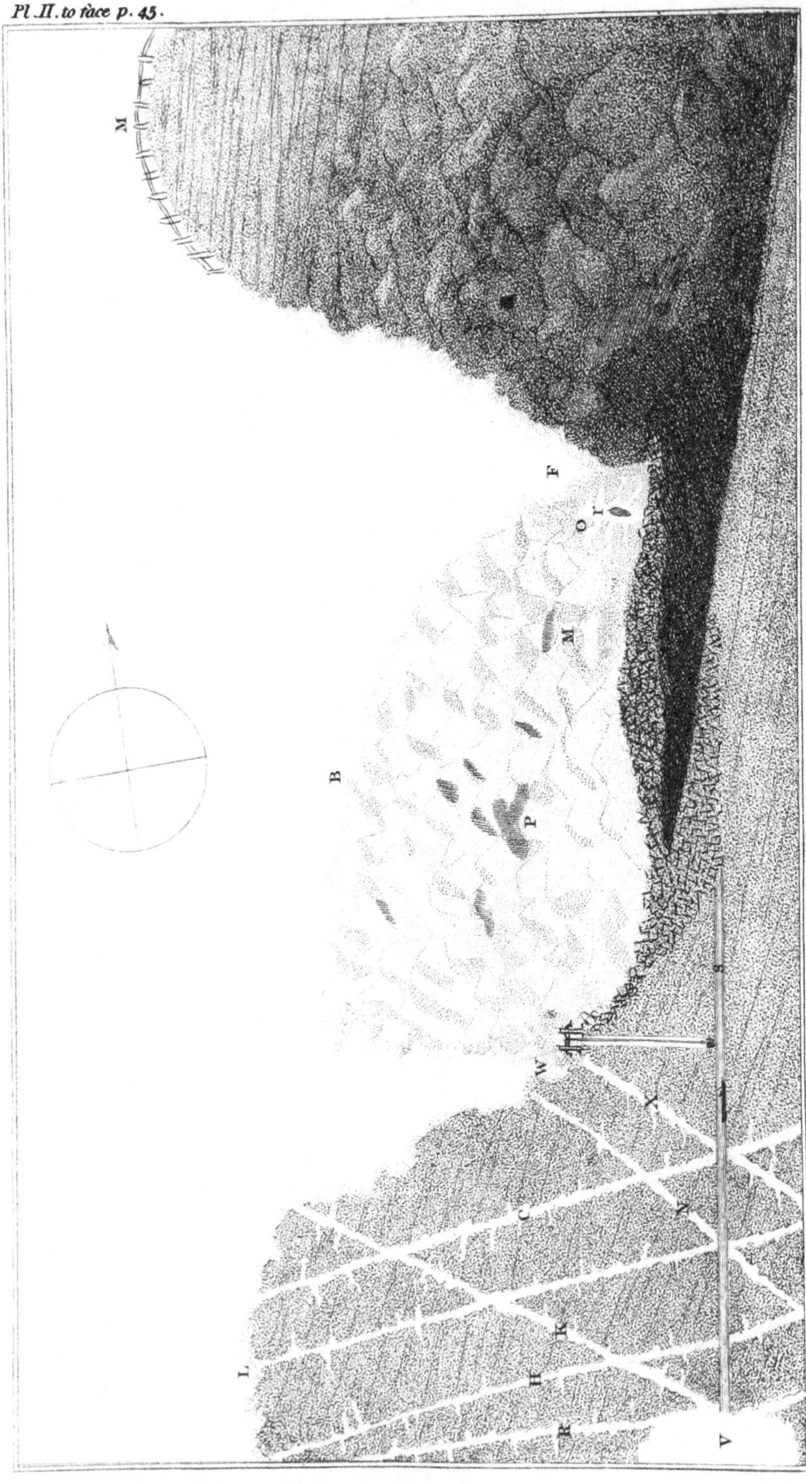

Mawe del.

H. Mutlow sc.

Published Jan.y 1. 1802, by J. Mawe, No. 5. Tavistock Street, Covent Garden, London.

SECTION IV.

Strata of the Mountains to the west of Castleton.

(See Plate 2.)

IF we suppose these mountains vertically divided, from the surface to the bottom, M will represent the mountain of Mam Tor. F is the fissure that separates it from the limestone mountain B, where the blue fluor is found. W is the fissure, where is the road called the Winnets; and L the mountains that compose the Long Cliff.*

Mam Tor, which is said to signify Mother Rock, presents on one side a bare stratum of 130 yards in perpendicular height, being composed of mircaceous' grit, in small

* Tablets of these strata may also be had, shewing the direction of the veins, caverns, &c.

and

and thin beds, intervened with ſhiſtus. The latter is acted on by the atmoſphere, and from its expoſed ſituation, ſoon decompoſes, and falls in great quantities, whence Mam Tor has been called the ſhivering mountain. The inclination is about one foot in fifteen to the ſouth.

The ſtratum of ſhale, or ſhiſtus makes its appearance underneath the grit; and at the north end of the mountain B is a vein of ore called Oden, a mine as repreſented at O. This is a long rake vein of lead ore, running from weſt to eaſt, and underlying or *hading* ſouth. It is ſaid to be a very ancient mine worked by the Saxons. The operations are conducted horizontally, the ore being cut out more than a mile from the entrance; in ſome places 60 yards below the level or horizontal entrance, and in ſome places as much above it. This vein is of various thickneſs, ſometimes eight

feet,

feet, at others not above four inches, when it is divided by a *rider*, as at *r*. Lead ore in great quantities, with many fine cryſtallizations of blende, barytes, fluor, calcareous ſpar, ſelenite, &c.

The entrance of this mine is in the limeſtone; and the ſtrength of the vein extends it into the ſhale, which it ſoon leaves, and then comes into the limeſtone again. The vein is in ſome places divided by the hard limeſtone called *rider*: in which caſe the miners, following the divided veins, work by each ſide of the rider, perhaps more than a hundred yards, till the veins again unite. The lead ore produces about 60 per cent; and the mine employs about 100 people, who chiefly reſide in Caſtleton, and are, in general, intelligent men. It is eaſy of acceſs, and the manager is always anxious to ſatisfy the curioſity of thoſe who wiſh to

viſit

viſit it, by rendering them every civility in his power.

Here is found that ſingular variety of lead ore, called *ſlickenſide.* This galena preſents a ſmooth ſurface, as if plated. Sometimes it forms the ſides of cavities, and on being pierced with the miners tool, rends with violence, and explodes with a crackling noiſe. The cauſe of this phenomenon has not been fully explained. I have ſeen a man, when he came out of the mine, only a few minutes after the explosion, who, regardleſs of the danger, had pierced the ſides of this ſubſtance, and was much hurt, and cut violently, as if ſtabbed about the neck and other places with a chiſel, whence he was unable to return to the mines for two weeks.

The ſection under the letter L repreſents part of a mountain, called the Long Cliff, forming awful rock ſcenery along the road

to

to Manchefter. This mountain is chiefly of limeftone; and that marked B has every appearance of having been feparated from it. Here is the Speedwell, or navigation mine, driven north and fouth, as at S, to cut acrofs the veins of ore, which generally run eaft and weft. It has not proved fuccefsful though excavated for half a mile, and connected with immenfe openings, as at V. Its waters are collected, and boats float from the entrance to the *Forefield.*

In this mountain are feveral fmall rake veins, containing lead ore, barytes, fluor, carbonates of lime, pyrites, rofe-coloured calcareous fpar, blende, &c. X is a rake vein running nearly fouth eaft by eaft, and north weft by weft; it is called Little Winfter, and there are feveral fhafts on it, which are above twenty fathoms deep. N reprefents Long Cliff rakes, on which there are a few fhafts.

H represents a small rake, or *scrin*, of lead ore, running nearly east and west. It ranges up the side of the mountain, and on it are a few shafts, thirty fathoms deep. It is worked open from the surface for a small distance. K is a rake vein larger than the others, called Faucet, or foreside rake, which has a direction south, by 67° 30' east, and north by 67° 30' west: it ranges from the top of the Long Cliff, to and across the Castle hill. A stratum of basalt and toadstone openly appears, about one hundred yards to the north of the rake K, at a place called Little Banks; and I regret that it cannot be represented in the section. R is a small rake, or cross vein, or scrin, called Rock Pipe; which, however, takes the usual direction of pipe veins. C is Long Cliff Pipe, a small rake scrin, but in a pipe direction. Many small veins cross the mountain, several of which are cut across by the

Speedwell

Speedwell mine, as may be ſeen in the plate.

The veins of ore in this mountain range under the toadſtone. On the ſurface of the limeſtone are frequently found quartzy cryſtals detached, ſome pyramidal with priſms. They are called Derbyſhire diamonds.

In the cave, or ravine, ſouth of the caſtle, on Cawler Hill, is an irregular baſaltic column, appearing like a detached maſs; and from it I have broken pieces containing jaſper, calcedoney, and quartz. The outſide is decompoſed. Adjoining is a ſtratum of toadſtone, which is alſo decompoſed; it appears like indurated clay, full of holes, with green globules, ſpar, &c. This ſtratum ranges to the eaſt and ſouth, and is of conſiderable extent.

SECTION V.

Account of the Adits or Galleries.

IN Derbyſhire there are many levels, adits, or galleries, to free the mines from water, which are often admired by foreigners. One of the moſt conſiderable is at Wirkſworth, called Cromford Sough, relieving an extenſive mineral tract of its water, to the depth of the drain. This *ſough* paſſes from the north eaſt to the ſouth weſt, and is full two miles in length. The adventurers, with a laudable ſpirit, expended 30,000 l. in its completion, and the mines pay a proportion of lead ore to the proprietors of this grand drain. Yet the mines about Wirkſworth

are

are now beneath the level; and it is rendered of no farther uſe.*

Another *ſough* has been driven from a lower level, that of the Derwent, at a great expence, and is called Wirkſworth moor ſough. It is to the eaſt of that town; and is near three miles in length. This level will lay the mines dry for ſeveral fathoms, but it is not yet ſettled what contribution is to be made from the miners to the proprietors. It is ſingular that a low level in the limeſtone lays a great courſe of country dry, all the waters falling into it for a conſiderable diſtance.

At Youlgrave is one of the longeſt levels

* The relieving of the mines at Wirkſworth, by the ſpirited enterprize of driving the level, is become only a ſecondary object; for the water delivered by it at Cromford, has proved of amazing value, and was the firſt ſtream employed by the late Sir R. Arkwright, to work his cotton mill. This water continues to work one of the largeſt cotton mills in the kingdom, and has the great advantages of not being ſubject either to conſiderable increaſe or diminution.

in this country, running from the Derwent to Alport, and called the Helcarr *fough.* The length is near four miles, and it coft upwards of 50,000l. It is driven through a variety of ftrata, and relieves a confiderable number of mines. At Stoke Hall is an adit, driven up to relieve the Edge Side mines, near Foolow, exceeding a mile and half in length.

In Derbyfhire there are many other levels, extending a confiderable length, but there are few fteam engines, except on the coal mines.

The manor of Caftleton has a royalty, called the King's Field, enjoying ancient and peculiar cuftoms and privileges. Any perfon who difcovers a vein of metal may take poffeffion of it, and upon application to the bar mafter, after proving that a fmall portion of lead ore has been obtained, a piece of ground is granted, and the mine

and

and its produce become the legal property of the discoverer, who then generally sinks a shaft, or takes the most easy method, conformable to the laws and customs, to excavate the vein, and bring the produce to the surface. The mass containing parts of lead ore, spar, &c. as cut from the vein, is called *bowse:* when drawn out of the mine it is broken small; the lead ore is separated from the sparry matter, by various operations, as washing, sifting, &c. and brought to a proper size; after which it is measured by the *Bar* master, who takes a certain quantity as *lot*, or duty, for the king, and for tythe. It is then conveyed to the furnace, where it is smelted into lead. The duke of Devonshire has a lease of the duty from the crown.

Calamine, blende, &c. &c. pay no duty; but they cannot be taken off the ground until the land owner be ſatisfied, he having the prior claim.

This county is extremely full of ſmall veins, almoſt every miner poſſeſſing more or leſs. Such are ſeldom rich in produce, and indeed they have been rarely analyſed; but were the produce of every new vein ſubmitted to analyſis, it might lead to many advantages.

SECTION VI.

Obſervations on Cat Dirt.

IT has already been obſerved in the end of Section IV. that there ſometimes occurs in the Derbyſhire mines a ſtratum of decompoſed toadſtone, with the appearance of indurated clay, full of holes, containing green globules, ſpar, &c. Having been informed that lead ore had ſometimes appeared in this ſubſtance, and afterwards ſeeing it mentioned by Werner (on the information of a Derbyſhire gentleman), and from him by Mr. Kirwan, in his Geological Eſſays, p. 288, I became anxious to diſcover the truth of this matter. This deſire was encreaſed by the recent work of Faujas de St. Fond, entitled Travels in England and Scotland,

Scotland, from which I ſhall beg leave to ſelect the moſt eſſential paſſages on this ſubject; particularly thoſe pages where he informs the reader that galena has been worked in the toadſtone ſtratum.

Fagus de St. Fond, p. 328, ſays, Toad-
'ſtone containing lead ore, Mr. Whitehurſt,
'and Mr. Ferber affirm, that in all the mines
'which have yet been opened, the vein of
'ore is found excluſively in the limeſtone,*
'and diſappears ſo completely on reaching
'the bed of toadſtone, that not the ſmalleſt
'veſtige of it is diſcoverable in the latter:
'but that on piercing through the toadſtone,
'however thick, the vein as certainly makes
'its reappearance; and this fact, they affirm,
'holds good through every vein of ſtrata,
'to any depth. This diſpoſition, however

* This is not exactly correct, as before ſtated in this work. It will be found that the lead ore is frequent in the ſhiſtus, and ſometimes in the coal.

'aſtoniſhing,

'aſtoniſhing, is in general true; and thence 'Mr. Whitehurſt conceived the opinion 'that the toadſtone which thus ſeparates the 'calcareous ſtrata, and interrupts the courſe 'of the ore, muſt be the reſult of different 'currents of lava. My thoughts on this 'ſubject have been already explained, but 'if there ſhould remain any doubt that the 'toadſtone is not a product of volcanic fire, 'the fact which I am now going to ſtate 'will be ſufficient to remove them.

'Doctor Pearſon having ſpoken to me, 'at Caſtleton, of a miner who ſold ſelect 'ſpecimens for the cabinet, we went to pay 'him a viſit. I purchaſed from him a col-'lection of the moſt intereſting minerals of 'Derbyſhire, and ſome fine pieces of fluor 'ſpar, the cryſtals of which were in the 'moſt perfect preſervation.

'In the courſe of converſation with him, 'I aſked whether it was true that no vein

'of

'of ore was ever found in the toadſtone?
'he replied, that ſuch had uniformly been
'the fact hereto, and though long employ-
'ed in the mining buſineſs, he had never
'heard that the ſlighteſt trace of lead ore
'had been diſcovered in that ſtone, but
'that he had juſt learned to his coſt, that
'the rule was not without exception, if not
'in reſpect to toadſtone, at leaſt as to the
'cat dirt or channel.

'On requeſting a further explanation, he
'told me he had been ruined by working,
'on his own account, a vein, which at firſt
'had the moſt promiſing appearance, but
'which, after opening a deep gallery, at a
'great expenſe, was loſt in a bed of chan-
'nel, where, however, it was again re-
'covered, but in too poor a ſtate to indem-
'nify him. As the mine was but a little
'way off, he offered to ſhew it to us,
'eſpecially when he perceived I doubted
'his

'his account: providing himſelf therefore 'with ſome mining implements, he deſired 'us to follow him, and we willingly com-'plied. We directed our ſteps about a mile 'to the eaſt of Caſtleton, along the ſteep 'ſide of a mountain which fronts it, and 'upon a narrow road about 200 feet above 'the ſubſequent plain. The mountain is cal-'careous; and in ſome parts exhibit traces 'of ſtrata, but its general diſpoſition preſents 'a uniform and continuous maſs, like moſt 'calcareous rocks of great elevation. Marine 'bodies are not very abundant here; I ob-'ſerved however a few fragments of en-'trochi, and ſome terebratula. Several 'lead mines have been opened in it, and it 'alſo affords calamine in an ochreous form.

'We ſoon reached the entrance of the 'gallery, which penetrates in an horizontal 'direction, and opens in the ſtratified part 'of the calcareous rock, in a ſeam of white

'calcareous

'calcareous ſpar, which preſents a ſmall but 'very diſtinct vein of galena, intermixed 'with fluor ſpar.

'This indication, which was regarded as 'very promiſing in a mountain which con-'tained ſeveral other lead mines, determined 'E. Pedley, and his aſſociates, to commence 'their operation; but ſcarcely had they 'reached the depth of twelve feet, when 'the limeſtone terminated, and they had 'the misfortune to meet with the channel. 'As till then there had never been any in-'ſtance of the moſt ſlender veins of metals 'being found in this unproductive ſtone, 'they would immediately have diſcontinued 'their labours, had not the ſame vein of 'galena, which they traced through the 'limeſtone, continued its courſe in the 'channel or trapp. This appearance was 'ſo extraordinary and novel, that, ſeduced 'by it, the miners purſued the ore in the

'channel

‘ channel to the horizontal depth of ninety
‘ feet, in the conſtant hope that the vein,
‘ which never exceeded an inch in thick-
‘ neſs, would ſoon enlarge its dimenſions.
‘ But the farther they proceeded, the trapp
‘ became ſo hard, and it required ſo much
‘ labour and expenſe to cut through it, that
‘ Elias Pedley told us he was on the point
‘ of altogether abandoning the work. This
‘ bed of trapp was little more than ſeven
‘ feet thick, but it is very probable it ex-
‘ tends a great way into the mountain, when
‘ it is conſidered that the gallery has already
‘ been carried ninety feet in an horizontal
‘ line, without diſcovering any appearance
‘ of alteration.

‘ This bed of channel, or cat dirt, is
‘ really a greeniſh trapp, very hard in the
‘ interior of the mine, but upon being taken
‘ out of the gallery, and expoſed for ſome
‘ time to the atmoſphere, it becomes friable,

‘ its

'its colour changes, and it paſſes into an
'earthy ſtate. It is probable that this de-
'compoſition ariſes from ſome inviſible
'particles of pyrites, which become effloreſ-
'cent, and cauſe the ſubſtance to fall into
'a detritus.

'Here then is a proof that galena has
'been found in a bed of channel, in which
'it has been traced in an uninterrupted line
'of 90 feet, accompanied with a ſmall por-
'tion of calcareous and fluor ſpar. This
'inſtance exhibits a direct and unequivocal
'exception to the obſervations hereto made
'reſpecting the mines of Derbyſhire. The
'exiſtence of lead ore in the trapp is a cer-
'tain proof that it is not the product of
'fire.

'I know that thoſe mineralogiſts who
'are converſant in the ſtudy of lithology,
'who have examined the trapp upon the
'ſpot, and are fully acquainted with that
ſtone

'ſtone and all its varieties have no occaſion
'for this proof. But the fact appeared of ſo
'much importance that I conceived it pro-
'per to mention it, to do away every
'doubt on the ſubject. This conſideration
'therefore, will form my excuſe to thoſe
'who may be diſpleaſed at the minute and
'tedious details which I have been obliged
'to enter into, that I might place the
'queſtion in the cleareſt poſſible point of
'view.'

I am ſorry Monſ. Faujas de St. Fond did not examine this ſubſtance more minutely, as well as the well-known mountain of Mam Tor.

He ſays, page 325 of his work, 'Several
'mines have been opened in the very ſteep
'calcareous mountain of Mam Tor.'

Its very appearance is the moſt oppoſite to calcareous mountains that can be conceived, and its component parts are micaceous grit, laminated with argillaceous ſhiſtus, and incumbent on the ſame ſtratum; as before mentioned.

It is by no means my wiſh to enter into the examination of the works of literary men with a view to confute them, I merely wiſh to ſtate facts as they appear, ſo as to preſent the ſcientific with authentic materials.

The excellent Lord Bacon introduced what is called the experimental philoſophy, in which facts alone are conſulted; and I hope to be pardoned, if ſacrificing for a moment my veneration for thoſe illuſtrious characters, I ſtate the plain facts with the freedom of a practical man addicted to no theory. I went into a mine called DIRTLOW, about a mile eaſt of Caſtleton, where it is ſaid that the vein of lead ore migrates into cat dirt, or toadſtone; and indeed the mine itſelf took its name from this cat dirt.

In

In a ſhaft, on the left of the road going to Bradwell, which proceeds from a large rake vein, I went down about 40 fathom. One ſide of the vein conſiſted of what the miners called *channel*, *cat dirt*, or *toadſtone*; and a part of the vein was full of that ſubſtance. I cut out ſome pieces myſelf, and directed others to be cut, all which I took with me. Upon examination, this ſubſtance was of a browniſh green colour, interſperſed with green earth, ſoft, and porous. It was by no means ſo hard as the generality of limeſtone, and appeared on the contrary to be in a ſtate of decompoſition. It efferveſced ſtrongly with acids, and on putting a piece in a heated crucible, I immediately perceived a ſtrong ſmell of ſulphur. In the dark it emitted a blue flame, and burnt to a dirty red. On applying it to the tongue, it was cauſtic, and greedily abſorbed moiſture.

It

It ſeemed to me to be a queſtion, whether this ſubſtance be not a limeſtone, ſtrongly impregnated with pyrites, which are in a decompoſing ſtate; the green earthy matter I ſuſpect to be chlorite.

At Pindar and on Tideſwell Moor, where the lead ore is alſo ſaid to occur in this ſubſtance, I examined another variety of it, but found it eſſentially to agree with the former. I therefore conceive that the confuſed terms of miners have miſled the very reſpectable authors before mentioned, who had not ſufficient time nor opportunity to inſtitute a ſtrict enquiry. In truth, the miners have applied the ſame name of toadſtone, or cat dirt, to ſubſtances extremely remote, and which have only a partial reſemblance in exterior appearance.

SECTION VII.

Account of the Fluor Mine, and of the manner of working that mineral.

IN the fourth ſection of this work, I have already explained the appearance of ſeveral Derbyſhire ſtrata. The reader will now forgive my proceeding to ſome, in which I am myſelf much intereſted. The mountain B, (ſee plate 2.) appears an aſſemblage of vaſt rocks of limeſtone, without connection or regularity, and is full of openings or caverns of immenſe depth, fiſſures, &c. In this mountain are the two mines that produce the beautiful compact fluor*, here called Blue John, which is found in *pipe*

* This ſubſtance acts as a ſpeedy flux to metals, owing to its peculiar acid, whence the name of *fluor*.

veins of various directions, as repreſented at P. In theſe mines it is neceſſary to arch the roads with ſtone; for after long rains, wood is not capable of ſuſtaining the weight. The fluor in various places appears to have been formed on the limeſtone; for it frequently has that ſubſtance for a nucleus, around which it ſeems firſt to have chryſtallized, till it had greatly increaſed by accumulation. Frequently, however, the centre is hollow.

In various parts of the mine, in caves filled with clay and looſe adventitious matter, the fluor appears in detached maſſes, bearing every appearance of having been broken from the limeſtone, on which it ſeems to have been originally formed; for every piece, in one part or other, ſeems to have adhered to ſomething, and to have been broken off These caverns are frequently

quently beſet with beautiful calcareous ſtalactites, of a large ſize.

It is impoſſible to account for the prodigious variety, and ſingular diſpoſition of the veins, and ſudden contraſts of the fineſt colours, which occur in this ſubſtance. Some of the pieces of fluor are a foot in thickneſs, and have four or five different and diſtinct veins; but ſuch large pieces are very rare. In general they are only about three or four inches thick; and ſome preſent one ſtrong vein, while others ſhew many ſmaller. Such as diſplay a geographical figure, like a coloured map, are moſt rare, and valuable. Some varieties are much more looſe in their texture than others. The colouring matter has been generally thought to be iron, but I ſuſpect it to be aſphalt, which may perhaps contain pyrites in a decompoſed ſtate; but there are many ſingular varieties which have not undergone any analyſis. The fluoric acid is easily obtained by pulverizing the fluor, and putting it in a leaden retort, to which add its weight of any of the mineral acids. Apply a gentle heat, and the fluoric acid will appear as gas, which may be caught in a vessel of the same materials with the retort. Its peculiar property of corroding glass and siliceous substances, is well known, and has been employed in France in engraving glass plates of singular beauty. It is also a noted flux for the lead ore, its very name being derived from its being so ready a mean of accelerating fusion.

Faujas de St. Fond has pronounced this substance to be the most beautiful in the mineral kingdom; and has particularly praised the elegance of the manufacture.

In the loose earth of the caverns are found rounded nodules of lead ore, sometimes called potatoe ore; and there is in the same mountain a pipe vein of calcareous spar,

one

one of which contained lead ore, which was worked as at M, called the Miller mine. The limeſtone that compoſes the whole is full of marine exuviæ. This mountain, as I before obſerved, reaches ſouthward to the Winnets, where it is ſeparated from the Long Cliff by a deep ravine, in which is the road to Mancheſter.

The rocks on the ſide of the road are ſtupendous, and in many places perpendicular, running in all directions, and forming immenſe caverns. The mines of this mountain afford the greateſt variety of mineralogical information of any which I have yet ſeen. The veins themſelves, the frequent obſtruction of their directions, and the diſlocation of the ſtrata, with the heterogenous ſubſtances found in the immenſe caverns, preſent matter for great ſtudy, and curious obſervation.

The acceſs into the mine of fluor is tolerably

ably eaſy, deſcending about 60 yards down ſteps, amid limeſtone. Proceeding about 30 yards deeper, by an eaſy route, you arrive at a moſt beautiful cavern, beſet with delicate white ſtalactite, which, to the imagination, aſſumes a variety of figures. At a ſmall diſtance further, you are led into a cavern yet more grand, in which ſome ſtalactites, hanging perpendicularly from the roof of the projecting rock, form a ſtriking ſemicircle; the black walls of the mine contraſt with the ſnow white ſtalactites, and conſtitute a ſcene ſurpaſſing deſcription. Hence you are led into a variety of intereſting caverns, veins, &c. and the guide will be ready to give every information to the curious viſitor, without any wiſh to delude him by fabulous wonders, or intereſted error.

I ſhall now proceed to give a ſhort account of the chief varieties of fluor, and of the

the method employed in their manufacture. Fluor, or *fluate of lime*, generally cryſtallizes in the cube and its modifications, rarely in the octaedral, and ſtill more rarely in the dodecaedral form. The chief varieties are the following:

Water coloured cryſtals of cubic fluor, ſtudded with bright pyrites. The accumulation of cryſtals frequently covers the pyrites with a pretty effect.

Very large and tranſparent cubes of fluor, with pyrites in the inſide, accompanied with blende and lead ore.

Blue fluor, of a violet colour, in perfect cubes, with cubes in the interior.

Amythiſtine and topazine fluors. The latter is of a fine yellow, with internal cryſtals of pyrites.

Dark blue fluor, with the edges bevelled on each ſide.

Blue

Blue fluor, with one bevelled edge, or a plane on each edge.

Blue fluor, with four fided pyramids on ch face.

Blue fluor, indented and perforated.

Fragments of octaedral fluor.

Ruby coloured fluor, in perfect cubes, on limeftone.

Granulated, or fandy fluor, of a rofe colour.

Compact fluor, in maffes, formed on limeftone, or in nodules. This feems an accumulation of cube upon cube, forming prifms, the furface of which is cryftallized. Some of thefe maffes, which are feven or eight inches thick, are feparated in two or three places with a very thin joint of clay, fcarcely thicker than paper. This variety is compofed of very fine veins, and fudden contrafts of blue.

Another variety in maffes, full of holes, containing

containing decompoſed calcareous ſpar, in the form of brown pearl ſpar. This variety is lightly veined with blue; and the bottom, or part next the rock, which is called the root, is wholly blue, and tranſparent, but not ſo dark or ſo finely figured as the veins.

Another variety, harder than the former, the ground clear white, but tinged like the *lichen geographicus*. This never forms veins.

A variety having five regular veins of fine blue. This ſtone is much looſer in its texture: and where cut acroſs its cryſtallization, it preſents a beautiful honeycomb appearance. There is another variety more regularly divided into three veins.

The dark blue, approaching to black, is perhaps of all others the moſt rich and beautiful, and diſplays a variety of pentagonal figures, and is bituminous.

The

The variety, which is of a dark purple, pervading the whole maſs, is looſe and friable.

That of one ſtrong blue vein is much harder, very rich, and tranſparent.

Fluor in detached cubes, in the limeſtone, appearing a little decompoſed.

Fluor with metalic veins.

Fluor decompoſing.

Fluor of a fine green tinge.

Of a blue colour, in a maſs of cryſtallized cubes, with elaſtic or indurated bitumen.

Fluor in compact limeſtone with galena, in veins and ſmall particles, filling up interſtices.

Fluor cryſtallized in cubes, upon hornſtone or petroſilex.

Fluor in the cavities of coralloids.

Fluor with barytes, commonly called tyger ſtone, being opake, and full of dirty brown ſpots.

Having

Having thus given an account of the chief varieties of fluor, I ſhall deſcribe the art of working it.

When it is intended to be worked into a vaſe, or the like article, a piece is ſelected fit for the purpoſe; and if after minute examination it be found free from defects, it is carved with a mallet and chiſſel into a ſpherical form, and then fixed on a *chock* with an exceedingly ſtrong cement. The chock is then ſcrewed on the lath, a ſlow motion is produced, and water continually drops on the ſtone, to keep the tool cold, which is at firſt applied with great care. This tool is a piece of the beſt ſteel, about two feet long, and half an inch ſquare: it is reduced to a point at each end, and tempered to ſuit the work. As the ſurface becomes ſmoother, the tool is applied more boldly, and the motion much quickened, till

till the piece of fluor be reduced to its intended form.

The laths worked by machinery poſſeſs a great advantage, the tool being applied with more delicacy, from the body not being in motion, as in turning the foot laths. Another great advantage is, that any motion is procured by a touch; as in ſome caſes a very quick motion is required, and in others very ſlow.

The piece being thus formed, and rendered ſmooth by the ſteel inſtruments, in order to render it fit to receive a poliſh, a coarſe ſtone is applied with water, ſo long as the ſmoothneſs is improved by theſe means. Then finer grit ſtone, pumice, &c. till the piece be ſufficiently ſmooth to receive coarſe emery, and afterwards fine emery.

If with the latter it appear of a good ſhining gloſs, then the fineſt putty is employed for a conſiderable length of time,

till

till the poliſh be as bright as poſſible, which is known by throwing water on it. If the part thus watered appear higher poliſhed than the reſt, the poliſhing is continued till water will not heighten the appearance.

The advantage of the lath, worked by water, is particularly conſpicuous in forming delicate hollow vaſes, &c. for by the uſe of the foot lath, the fluor was frequently broken, and its laminated texture at all times diſturbed; but the uſe of the water lath, by its ſteadineſs, prevents theſe inconveniences.

The firſt mill that was built for Sir Thomas Lombe, at Derby, is now converted into a manufactory for this purpoſe, as mentioned in the firſt ſection. This beautiful production of nature is here formed into elegant urns, vaſes, columns, &c. giving employment to a number of families, and forming an intereſting article of commerce.*

* Meſſrs. Brown and Co. the proprietors, are happy to ſhew travellers their manufactory, and give them every information. Their wholeſale warehouſe in Taviſtock-ſtreet, Govent-garden, exhibits the greateſt variety of elegant urns, vaſes, &c. formed of this beautiful ſtone, at the ſame price as at the manufactory; alſo the moſt ſplendid and extenſive collection of minerals in the kingdom.

SECTION VIII.

Account of other Minerals found in Derbyshire.

THE gypſum or alabaſter, naturally ariſes to obſervation, after the fluor, as being employed in works of ſimilar elegance. This ſubſtance is found in large maſſes, filling up cavities or inſulated places in the argillaceous grit, near Derby; as at Elvaſton, Chellaſton, and Tetbury. It never forms a ſtratum, but is generally attended with gravel, ſtrong red clay, and an earthy covering, in which are frequently innumerable ſhells.

The gypſum is generally veined with red, and frequently with a dirty blue. No mineral, or marine ſubſtances, are found in it.

 Gypſum

Gypſum or alabaſter, is generally ſo ſoft as to be ſcraped with the nail; but ſome ſorts are much harder than others, and of a cloſer texture. Near the ſurface it is ſtriated and ſometimes cryſtallized; below it is much more compact, and is frequently uſed for architectural purpoſes, forming elegant columns, pilaſtres, &c. of which thoſe in the hall of Lord Scarſdales, at Kedleſton, ſtand unrivalled. When cryſtallized it is called ſelenite. It is eaſily calcined, and then forms what is called plaiſter of Paris, which greedily abſorbs water, and is caſt into various figures, as imitations of the antique ſtatutes, &c. It is likewiſe uſed for moulds, for floors in houſes and other economical purpoſes. It forms an article of trade, and conſiderable quantities are ſent to London.

The chief varieties of this ſubſtance are capillary gypſum, in delicate ſilky filaments, three or four inches long, ſo tender as to render it impoſſible to procure it perfect.

Plumoſe

Plumoſe gypſum, like white feathers, elegantly curled, on limeſtone.

Green ſelenite, extremely rare.

Selenite in tranſparent priſms and rhombs.

Gypſum, rock alabaſter. Striated ſilky alabaſter.

Compact white; ſemi-tranſparent; red veined; variegated, &c.

I ſhall now give a ſhort detail of the other minerals and metals found in Derbyſhire.

In the ſiliceous order may firſt be mentioned topazine and roſe coloured quartz, in hexagonal priſms, with double pyramids detached.

Amethiſtine quartz finely tinged; with perfect hexagonal priſms, alſo with double pyramids detached.

Clear colourleſs quartz in fragments, and the ſame encloſing bitumen: theſe varieties are looſe in the limeſtone.

 Chert,

Chert, hornſtone, or petroſilex, forming thin luminated beds, near Bradwell, Buxton, Middleton, &c. &c.*

The ſame ſubſtance is alſo found exhibiting entrochi, coralloids, &c. in which caſe it ſeems the ſecondary petroſilex of Sauſſure.

Of the barytic order the moſt general is the ſubſtance called cawk, from its reſembling chalk, (which is not found in the north.) It occurs in great quantities, being the common attendant on lead ore. The colour is often white, but more frequently a greyiſh white, inclining to the cream

* Dr. Smith in his travels, vol. I. p. 176, mentions a ſtratum of flint running horizontally through the limeſtone by the rock houſe, at Cromford near Matlock. Mr. Kirwan in his geological eſſays ſays, that it is found in ſtrata 12 feet thick in Derbyſhire. For this he quotes the philoſophical tranſactions. In Peak foreſt are a variety of chert beds of various thickneſſes, ſome are in contact with the granulated limeſtone, although limeſtone full of ſhells is above it and below it; its colour is of the dove blue, it adheres to the chert, and is ſofter than the other variety.

tinge,

tinge, which ſometimes riſes to the ochre yellow. It is ſoft but ponderous: fracture earthy, ſometimes ſcaly. It often contains ſmall veins of lead ore, as thin as threads; and ſometimes ſmall veins of fluor and blende.

Barytes occaſionally occurs cryſtallized in tabulated rhombs, on grit ſtone; but more generally in delicate tabulated cryſtals, which by combination, form ſpherical balls. One variety is ſtalactitic, ſometimes with tranſparent cryſtals, and native ſulphur.

The arboreſcent barytes is compoſed of ligaments of various colours, interveining each other, appearing ſomewhat like branches with foliage. A piece now before me is poliſhed, and exhibits dark brown and lilac figures, beautifully interſperſed with blue in a geographic form, or like a coloured map, and affording beautiful contraſts.

Barytes in tabulated cryſtals, opake white, half

half an inch in diameter, but as thin as leaf gold, on a cellular gypſeous matrix, with native ſulphur.

Barytes having a plumoſe appearance, when covered with tranſparent cryſtals of fluor. Barytes in fluor forms a pretty variety.

Barytes has lately been found confuſedly cryſtallized, of a ſky blue colour: the fracture foliated.

But what the chymiſts call carbonates of lime, and mineralogiſts calcareous ſpar, &c. afford an amazing variety of colours and cryſtallizations. This ſubſtance is apt to be confounded with fluor, from which it eſſentially differs; the fluoric acid being of a peculiar nature, and very different from the carbonic, not to mention other diſtinctions. The calcareous ſpar here appears in its moſt uſual ſhape of the rhomb, and its modifications, macles, &c.

The

The primitive rhomb is rarely found. It is generally on a dark bituminous limeſtone with pearl ſpar and ſelenite: the primitive rhomb paſſing into a variety of modifications.

Lenticular cryſtals, on dark limeſtone, blende, &c.

The dogs tooth ſpar, forming double hexagonal pyramids, joined at the baſis.

Hexagonal cryſtals of calcareous ſpar, rarely terminating with pyramids of the primitive rhomb.

Hexagonal cryſtals terminating with the primitive rhomb truncated.

Hexagonal cryſtals terminating with the lenticular pyramid.

Hexagonal cryſtals with a variety of terminations forming pyramids, with three, ſix, twelve, fifteen, and more facets.

Hexagonal priſms of a high topaz colour, with various terminations.

Fibrous,

Fibrous calcareous ſpar. Calcareous ſpar appearing mamellated.

Macles, or twin cryſtals; ſome exceedingly rare, and in great variety.

Opake ſnow white calcareous ſpar, cryſtallized in double hexagonal pyramids, joined at their baſes.

Stalactites forming a variety of beautiful colours, with the appearance of agate vein.

Stalactites, the terminations cryſtallized.

Green ſtalactites, very rare.

Granulated calcareous ſpar, or in maſſes, compoſed of grains.

Roſe coloured calcareous ſpar, amorphous.

To this order alſo belongs a great variety of marbles. The upper ſurface of the limeſtone is frequently nearly white, probably from being bleached by the weather; it is perfectly hard; a variety of it is found at Lover's Leap, near Buxton.

Of

Of a fine ſtone colour, near Hopton.

Fine red brown, near Aſhford, full of large marine remains.

Fine black is found at Aſhford, Matlock, Monſaldale, and various places.

Coralloid marbles exhibiting a variety of madrepores, are found in laminæ in various parts of the ſtratum.

Dark coloured limeſtone, full of marine exuviæ, in large figures.

Dark coloured limeſtone in very minute figures at Wetton. Frequently in the ſtratum of limeſtone are decompoſing green globules, and ſmall green veins with decompoſing pirites. This ſubſtance is ſoft, and of a fine light green colour.

Before proceeding to the metallic ores, a few of the inflammables may be mentioned. Among theſe the moſt peculiar and remarkable is the elaſtic bitumen, in its various ſtates, (or mineral cahoutchou) a recent

diſcovery.

difcovery. It is generally found between the ftratum of fhiftus and the limeftone, rarely in fmall cavities adhering to the *gangart*, and fometimes containing lead-ore, fluor, &c. When firft detached the tafte is very ftyptic, as if blended with decompofed pyrites. It varies in colour from the blackifh or greenifh brown to the light red brown, and is eafily compreffed ; but fometimes the fame piece is lefs elaftic in one part than in another. On burning it the fmell is rather pleafant.

A piece of the elaftic bitumen, of a reddifh brown colour, now before me, contains nodules of indurated fhining black bitumen, refembling jet. This kind is very rare.

Another variety, the only piece I have feen, is in a *marine fhell*, in a piece of limeftone.

The elaftic bitumen of a dull red, and transparent,

tranſparent, in cryſtallized fluor, extremely rare.

A variety, yet more ſcarce, but leſs elaſtic, appears to be compoſed of filaments, and has a ſingular acid taſte. The characteriſtics are very different from any other ſort; and might probably, if inveſtigated, account for the origin of this ſubſtance. On cutting, and in other circumſtances, it reſembles ſoft cawk, or old bark from a tan-yard.

Indurated bitumen, appearing like jet, in amorphous maſſes, and globules of a ſhining black, but ſometimes liver-coloured. This kind is electric when rubbed; and is ſometimes found in barytes.

Elaſtic bitumen with aſphalt, containing lead ore. The ſame in long filaments, almoſt as fine as wire.

Sulphur (native) in the cellular parts of baroſelenite on limeſtone.

That

That eſſential mineral, Coal, which gives birth and ſupport to many of our manufactures, appears in different parts of this country. It is found in the greateſt plenty in the north eaſt, as has been before mentioned. Towards the north weſt of Derbyſhire it is found again near Buxton, as at Coit Moſs, on the edge of Cheſhire. In various places, for the diſtance of ſeveral miles from eaſt to weſt, neither the ſtratum of argillaceous grit, nor the coal, have appeared.

Coit Moſs is a conſiderable mountain 3 or 4 miles weſt of Buxton, compoſed of argillaceous grit; at the depth of 30 or 40 fathoms, beds of coal are found.

There appears a feruginous ſhiſtus-like ſubſtance, ſeveral feet thick, incumbent on the coal, which decompoſes by expoſure; its baſis is argil with oxyde of iron.

The

The coal is a great relief to this mountainous and cold country, and its effects are conspicuously seen in a variety of objects.

Sulphur combined with Iron, or Martial Pirites.

IN enumerating the metallic ores, I shall first mention iron, which appears in the form of sulphuret or pyrites in various states, but generally crystallized in the octaedron, cube, dodecahedron, &c. is frequently mamelated, elliptical, arborescent, and kidney form; colour shining bright yellow, sometimes inclining to brown; very brittle; gives fire freely, and when conflicted has a very sulphureous smell; it is frequently compact, forming a vein.

Hematites,

Hematites, or liver ſtone, is ſometimes found incumbent on a ſolid maſs or ball; pyrites, about a quarter of an inch in thickneſs, the interior braſs yellow, diverg-ing from a centre.

The argillaceous iron ore is in the moſt general uſe in the iron works. It forms a thin ſtratum in the coal countries, ſome-times encloſing ſhells and coralloids. It frequently occurs in nodules above coal, containing vegetable impreſſions. It is ſome-times mixed with a proportion of Lanca-ſhire ore; which by uſing a proper quan-tity of limeſtone as a flux, is found to be a conſiderable improvement.

Calcareous or ſparry iron ores, are of a fine browniſh red colour, ſometimes bright yellow, ſcaley, and dirty brown, found in amorphus maſſes near the ſurface, and fil-ling inſulated places. The calcareous matter

ſeems

ſeems predominant, the cryſtallization is frequently preſerved, and it appears in different ſtages of decompoſition; it is very uſeful to mix with other iron ores, and is ſaid to make a good iron for converting into ſteel.

Manganeſe appears in the ſhape of *black wad*; formerly ſuppoſed to be an iron ore, in earthy maſſes, crumbling to powder on expoſure to the atmoſphere, being very looſe and friable. Theſe black lumps are not unlike hard balls of ſoot, but when broken, capillary veins appear ſomewhat of a metallic luſtre.

Black wad muſt not be confounded with *black jack*, which is a blende or ſulphuret of zinc, or pſeudo galena, found in amorphous maſſes, frequently cryſtallized, and generally accompanying fluor and barites. The colour is a blackiſh brown, inclining to a metallic luſtre, and a little tranſparent.

 A variety,

A variety, called ruby blende, is cryſtallized on calcareous ſpar, and is of a beautiful tranſparent red.

Another variety is called pigeon necked blende, from its irideſcent hues.

Red blende, minutely cryſtallized on fluor.

The blendes generally produce above forty per cent of zinc, ſometimes with iron, lead, or copper, and are frequently in Hungary and other places, auriferous.

Zinc* is alſo found in the form of lapis calaminaris. This native oxyd occurs of various colours, brown, reddiſh, and blueiſh brown, yellow, waxy, green, white, ſtalactitic, porous, &c. It is found in nodules, and often clothes calcareous ſpar, which it

* Native zinc is ſaid to have occured once: but after making every inquiry, and not having been able to ſee the ſpecimen, I feel warranted in concluding, from an intimate acquaintance with the mine which is ſaid to have produced it, that it is improbable, and wholly a miſtake.

ſoon

ſoon decompoſes. I have ſeen *fragments* of calcareous ſpar, coated with calamine: a ſufficient proof of the recent formation of the latter.

It is ſometimes in an ochreous ſtate, combined with ferruginous matter, but the compact is the beſt; and it is moſt eſteemed when of a waxy colour. Sometimes there are tranſparent tabulated cryſtals, and it is frequently botroidal, or in the form of grapes, and alſo ſtalactitic. The ſnow white is mamellated, and is extremely rare. It ſeldom occurs coating fluor, but often decompoſing calcareous ſpar, and is frequently attendant on blende.

Calamine generally contains *above* ſixty per cent of zinc, with ſome iron; and aſſumes various appearances, ſometimes efferveſcing with acids, and phoſphoreſcent.

To the eaſt of Caſtleton is the place called Red Seats, where are ſeveral rake veins containing blende, calamine, and ſmall quantities of lead ore, with barytes, calcareous ſpar, and fluors. Here are ſeveral ſhafts of no great depth. In the vicinity are ſeveral maſſes of limeſtone, conſiſting of ſmall pieces, or angular fragments, cemented by the water filtering through the maſs, and precipitating its earthy particles, which conſtitute a ſtalactitic matter pervading the interſtices.

SECTION IX.

Of the Lead Ores.

THOUGH lead ores generally contain ſilver, none in Derbyſhire, yet analyſed, yield any portion of that precious metal, ſufficient to defray the expences neceſſarily attendant on the ſeparation of it.

The moſt common lead ore is galena, or ſulphuret of lead, which generally lies in larger or ſmaller veins and maſſes; frequently in nodules, with cawk, a name here uſed for barytes. Galena is frequently cryſtallized in cubes, with the angles truncated, alſo in the octaedron and its modifications. It is of a bright luſtre, and flaky fracture.

fracture. Another variety, when broken, is remarkably bright and foliated ; by expoſure it becomes tarniſhed and decompoſes.

Another kind of galena is called the ſteel grained lead ore; being very hard, and the granulated appearance, when broken, reſembling the fracture of ſteel. This ore ſometimes appears fibrous, not unlike the common compact ore of antimony.

Maſſes of galena frequently contain ſmall holes, the ſurfaces of which, being nearly black, appear as if corroded. Sometimes carbonate of lead appears on it, in various ſtates and forms ; ſome of the cryſtals having a ſemi-metallic appearance, others of a dirty white, and ſome tranſparent : the ſhape is chiefly the priſm, and the double hexagonal pyramids joined at the baſe.

Two, three, or four veins of galena ſometimes occur in barytes, the whole not broader than two and a half inches. Theſe

veins

veins are perpendicular, and afford a pleafing image of the large veins of ore.

Spherical nodules of lead ore are not unfrequently found in caverns in the mines, whither they muft have been conveyed by water. Some of them are hollow, and contain native fulphur.

A pulverulent black lead ore, fometimes diffeminated on the matrix, appears to arife from the decompofition of the galena, owing probably to their fuper oxygenation.

Slickenfide is a fingular variety of galena, appearing of a bright metallic luftre, with a reflection approaching to that of a mirror. It is thin, as if it only plated on one fide of a fubftance called *kevel;* and forms the fide of a vein or of a cavity. When firft pierced it cracks and flies with violence, as already mentioned. A new variety of flickenfide of a metallic luftre, coated with blende of a light ftone colour, fometimes dark brown on a fine violet fluor matrix, has been recently found.

Maſſes of lead, perfectly malleable, but very much coroded, are ſometimes, though very rarely, found in old mines. They appear ſtalactitic. At an early period the miners made fires in the mines to melt the lead ore in the veins, and this ſubſtance may probably have remained there ever ſince.

The antimoniated lead ore runs, like net work, in filaments curiouſly interwoven, and is ſometimes accompanied with indurated bitumen. This kind is rarely iridescent.

A moſt beautiful irideſcent variety is ſometimes met with in octaedrons, the colours being very vivid at firſt; but they are ſubject to tarniſh and looſe their beauty, by expoſure to the atmoſphere. This variety is generally attended with cryſtallized fluor, affixed to its ſurface.

Sometimes

Sometimes a variety of carbonate of lead occurs, which does not adhere to the galena. Maſſes have been found of a horn colour, ſemi tranſparent, and finely cryſtallized on the ſurface.

Muriate of lead in perfect cryſtals of a beautiful tranſparent yellow colour.

What is called glaſs lead, *appears* as if it had undergone the action of fire; is tranſparent, and ſometimes cryſtallized; but in other inſtances is of an opake, waxy white. It is eaſily melted by the blow pipe.

Nodules of carbonated lead have alſo been found, formed by a combination of priſms, acicular, fibrous, and interwoven, ſometimes of a conſiderable ſize in looſe earth. Other carbonated nodules found in a looſe ferruginous earth, granular, and of a ſhining micaceous fracture, and eaſily reduced

duced to a ſandy powder. This variety may be termed ſandy lead ore.

Cryſtals appearing ſemi-metallic: ſometimes one part of the ſame cryſtal a dull blue colour, the other tranſparent horn colour.

A ſingular variety of carbonated lead occurs in ferruginous earth, in nodules, with hydrophanous ſteatite, &c. appearing like a decompoſed breccia, in ſmall ſemi-tranſparent veins.

Theſe ores have hitherto attracted little notice; nor indeed, till within theſe few years, was it known that they contained lead.

An argillaceous variety, called wheat ſtone, is found in a large vein. It is of a light ſtone colour, very heavy, with black ſpots, and contains arſenic. It is not tranſparent: fracture earthy, with a few bright

metallic

metallic ſcales, and ſometimes traces of ſmall ſemi-metallic veins. This variety is extremely eaſy of fuſion, during which it emits a ſtrong ſmell of ſulphur and arſenic.

Phoſphate of lead, of a leek green colour, in hexagonal priſms, is ſometimes found on barytes, attached to ſand-ſtone.

Molybdate of lead, of a fine yellow colour, approaching to orange, ſometimes appears in the cavities of galena, and of carbonated lead. This variety I have ſeldom met with in this county.

Galena generally yields from 50 to 80 per cent. at the furnace; many arts are practiſed in the dreſſing, to make it appear clean and rich, in order to fetch a higher price, which are well known to the ſmelters or ore buyers.

The carbonates of lead are ſo full of heterogeneous matter, that they rarely

yield

yield more than from 30 to 50 per cent. and do not produce ſuch ductile metal as the galena.

To face p. 109.

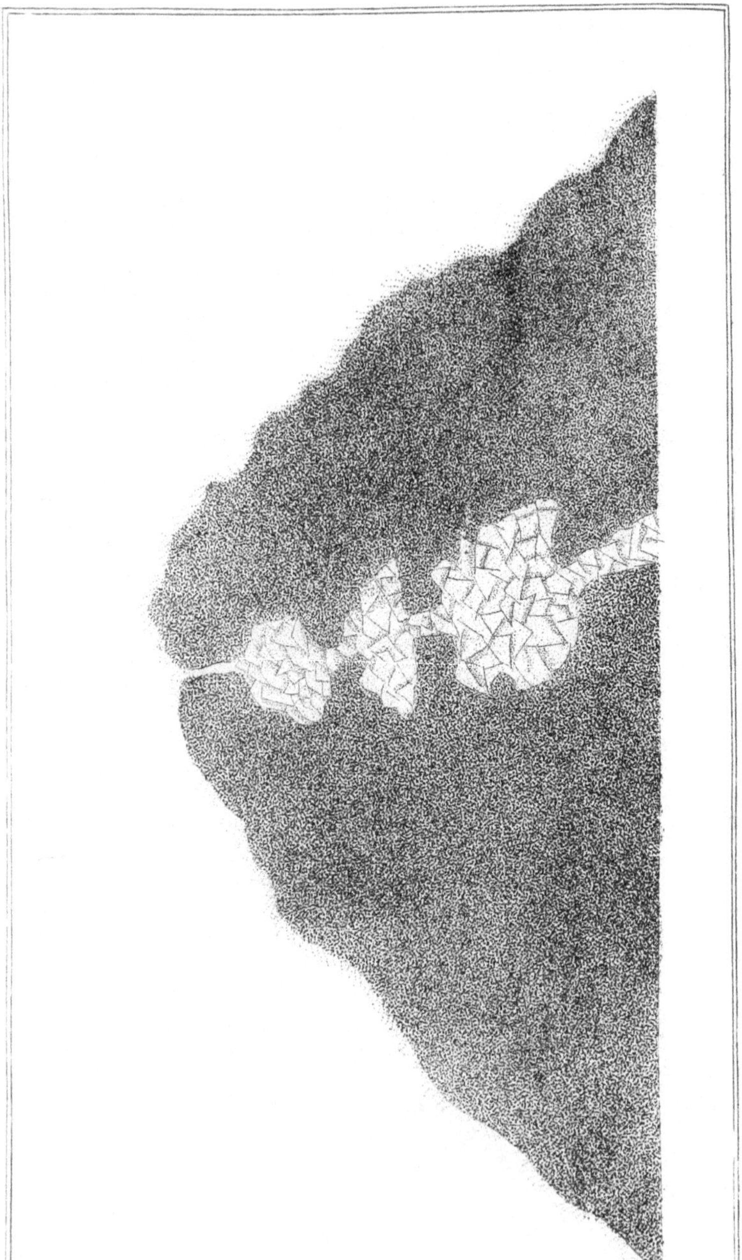

SECTION X.

Account of the Ecton Copper Mine.

THE Ecton copper mine is the only one of any consequence in Derbyshire; to which though on the edge of Staffordshire, it is generally reputed to belong.

The general produce of this mine, is massive rich yellow copper ore, frequently in contact with galena and blende, but specimens occur of purple, steel-blue, brown, or brass-yellow colours. The ore yields from forty to sixty per cent. and is sometimes vitreous and black.

Sometimes, though rarely, it is crystallized in the cube, and its modifications.

No specimens whatever can exceed the beauty of some from this mine, consisting

of

of iridescent copper pyrites, on a white barytic gangart. The colours are beyond description; the topaz yellow and gold; the violet and azure, being blended in the brightest effulgence.

The calcareous spar of Ecton, is a singular modification of the rhomb, very transparent, sometimes of a rich topaz colour, and generally containing brilliant crystallized pyrites in the interior.

Fluor, water coloured or light blue, also appears, finely crystallized with galena.

By the decomposition of the copper pyrites on the calcareous spar, arises a beautiful green efflorescence, clothing the spar, and sometimes appearing to pass into pearl spar.

Ecton also produces mountain blue, and mountain green; the former approaching to azure, the latter to a light verdegris colour, the fracture of these substances is earthy

earthy and uneven. They abſorb moiſture, and *appear* to be compoſed of barytes, granulated calcareous ſpar, and clay, with iron, and green calx of copper.

The famous vein of copper ore called Ecton mine, lies near Hartington, being what the Germans call a *ſtock work*, and the only one in this kingdom. It is ſituated from the ſurface to the bottom in a blackiſh brown limeſtone, the ſtrata of which are in the greateſt confuſion, extremely irregular, and running in all directions, as the reader may judge from the annexed plate.

This mine was probably worked at a very early period; it is one of the deepeſt in Europe, and it is now worked to the depth of 220 fathoms or 1320 feet; during the time it produced the greateſt quantity of ore, the profits where immenſe.

This work ſeems very different from the generality of veins; it has the appearance of of large cavities or openings in the ſtratum filled with copper ores, &c.

There are ſome few other mines in the neighbourhood of little conſequence.

This mine was extremely productive, and at one time employed more than 1000 people; the rich ore was in amazing large heaps, being in ſome places 70 yards broad, in others not above ten. It was ſmelted at Cheadle, where coals are more plentiful; and the copper is greatly eſteemed, and much in requeſt for large boilers, &c. being more ductile than any other.

SECTION XI.

Deſcription of the Surface of the Country in Derbyſhire.

AFTER having premiſed the ſtrata, it perhaps may not prove unintereſting to give a ſhort deſcription of the ſurface of this county. The ſubſtances found in each ſtratum have been before mentioned, nor have we any account on record, or roof, of other ſubſtances having been found in them.

To begin with that part of the country where the argillaceous grit appears at the ſurface, or under the vegetable earth, will be the moſt regular method; as it is confidered the uppermoft ftratum, and for a confiderable diftance prefents more uniformity; the hills are more regular, and rife by eafy inclination, forming vales of confiderable extent. The foil above is generally inclinable to the red clay and vegetable earth, where it continues unimproved. A large tract of country around Derby, in a high ftate of cultivation, has led me to the obfervation. In this neighbourhood, the value of coal is moft confpicuoufly feen in agriculture.

The effects of lime on thefe lands is tolerably afcertained, but in what manner it acts, has not perhaps been thoroughly examined. Incapable as I feel myfelf to inveftigate this fubject, yet, if by any means I could contribute to the examination of one fo interefting, and of fo great public utility, I fhould confider myfelf as not doing my duty, did I omit to mention my ideas.

The

The varieties of limeſtone render it highly neceſſary, that its properties and different characters ſhould be more generally known. Some ſorts are more proper for the purpoſes of agriculture, while others claim the merit for architecture. Limeſtone containing manganeſe, iron pyrites, and earth of the magneſian genus, is deſtructive to vegetation, according to the proportion it contains, but theſe ſubſtances do not render it unfit for a cement. The lime that contains the largeſt portion of carbon, and free from metallic ſubſtances, is conſidered moſt proper to ſtimulate and increaſe vegetation. Lime on clay lands, probably acts as an abſorbent, the vitriolic acid, which iron generally imparts to it, is in part diſengaged; by which means a ſubſtance deſtructive to vegetation is deſtroyed. Lime alſo acts powerfully, by preventing large maſſes of earth from forming by adheſion, and renders theſe maſſes more friable where it enters, by filtering, as it were, through the ſoil at the ſurface. It may be of conſiderable uſe by ſo greedily abſorbing moiſture, dividing the earthy particles, and forming a thin ſtratum a few inches below: and having regained a conſiderable part of the carbon, which was diſengaged by burning, it probably imparts it to the young plants.*

To the learned Doctor Darwin, the public are much indebted for information on this head; this truly great philoſopher has clearly ſhewn, that carbon is the life of the vegetable creation. I hope the reader will pardon me for the digreſſion, and truſt my motive will prove a ſufficient excuſe.

* In many of the ſparry limeſtones, that have been ſubmitted to my examination, I found a conſiderable portion of the phoſphoric acid, which may probably act as a great ſtimulant.

The

The face of the country, where the coarſe gritſtone makes its appearance on the ſurface, next ſtrikes the attention.

This ſubſtance forms long and narrow mountains rather than hills; the ſoil above it, in moſt of the elevated ſituations, as the Eaſt Moor and Edgeſide Hills in the neighbourhood of Caſtleton, Winhill, Lords Seats and others in the north, is very indifferent. On Cromford Moor it forms rude ſcenery, but more particularly ſo near Birchover, where immenſe maſſes lie in the rudeſt directions.

The mountains formed of the grit ſtone are the higheſt in the county, and have little depth of ſoil; but the land immediately below has more, and pays the cultivator for his improvement: decompoſed vegetable matter forms the beſt ſoil, and is waſhed down by heavy rains. A large tract of country might be mentioned where this ſtratum appears uppermoſt, but as it has the general character before deſcribed, it will be unneceſſary.

The ſurface where the ſtratum of ſhale or ſhiſtus makes its appearance is next in ſucceſſion. It moſt frequently appears uppermoſt in vallies formed by limeſtone mountains on one ſide, and grit ſtone on the other, where it is generally covered with looſe irregular pieces of ſtoney matter, called ratchell, which has probably fallen from the adjoining mountains in the lapſe of ages. Shale is ſubject to decompoſe by the action of the atmoſphere, and where it is expoſed, it falls into a black earthy matter; it is not conſidered as a ſubſtance friendly to vegetation, though when immediately in contact with limeſtone, its properties appear altered. In many places it is much impregnated with vitriol and martial pyrites. Lime acts very powerfully on it, and in many vallies it is

well

well covered with vegetable earth, and forms good land.

The form and general appearance of limeſtone mountains next preſent themſelves to view. In many parts they are perpendicular and overhanging, preſenting bare rocks in a great variety of forms, with diſtinct marks, ſtratification, openings, or caverns, of which none of the preceding ſhew any character.

The appearances of diſlocation and ſeparation in theſe mountains, are evident marks of the violent efforts of nature. Limeſtone in the north and weſt, in this county, generally forms large tracts of mountains, riſing to a conſiderable height from the valley to the ſummit; they then range more regularly to their furtheſt extent.

The lime generally uſed in the ſouth of Derbyſhire, is brought from Breedon, on edge of Leiceſterſhire, near Ticknall, and probably from ſome other places adjacent. North of Derby large mountains are formed in the neighbourhood of Wirkſworth, Cromford, Matlock, Winſter, Aſhford, the banks of the river Wye, a large tract from Aſhford to Buxton, in which is the beautiful valley of Monſaldale, Chee Tor, and a variety of places that preſent as fine rock ſcenery as almoſt any country can boaſt of, not forgetting in the more weſtern part, the beautiful valley of Dovedale, where the rocks are ſingularly picturesque.

The wild ſcenery of Middleton dale, and the Winnets on approaching Caſtleton from the north, is the admiration of viſitors, and the irregularity is beyond deſcription. The mountains which form one ſide of the beautiful valley of Caſtleton, are called, Longcliff, Cawler, and the Red Seats: they are full of veins of lead ore, and range

from

ſrom thence, ſouth, 8 or 10 miles; ſouth-weſt of Caſtleton, the limeſtone mountains range to Buxton, over Peak Foreſt, where is produced the fineſt lime.

The ſoil above this ſtratum is converted to all the purpoſes of agriculture; it affords moſt excellent paſture and graſs lands, and produces fine crops of corn. The mountains are uſually ſtored with cattle, being inconvenient for the purpoſe of tillage. Their ſtratification particularly engaged the late Mr. Whitehurſt's attention, eſpecially about the neighbourhood of Matlock, where he ſays (according to his ideas of the formation of the earth) they take an undulating form for a conſiderable diſtance, in which I conceive he was miſled, by not more cloſely examining the ſubſtance. The ſtratification of the high Tor and adjoining mountains, inclines into the rock or to the eaſt, probably as much as one foot in ſix, forming

forming an angle of 25 degrees. The face of theſe rocks is full of hollows, and very uneven, by parts projecting, which cauſes it when viewed at a diſtance to have an undulating appearance.

It will obviouſly ſtrike the reader, that if a prominent part is marked by ſtratification, and a part annexed recedes twelve feet, it will appear when viewed in front at a diſtance, that the ſtratification of the projecting part, is two feet higher than the part which recedes. It in reality is ſo, therefore it is the uneven form of the mountain, that gives it the undulating character, which it does not poſſeſs.

There is one place exactly oppoſite Matlock bridge, that ſeems to have more of the undulating form; and though I convinced my friends of the error of their opinion in the mountain called the High Tor, yet they ſeemed confident I ſhould agree with them

them in this inſtance, which I ſhall briefly deſcribe.

The front of a bare rock preſents itſelf of no conſiderable ſize; its form is broad at the bottom, narrowing to the ſummit: and the marks of ſtratification, riſing from each ſide, meet in the centre, forming a ſhort undulating appearance. This ſeemed to me contrary to any formation of limeſtone I had yet ſeen, and made me determine to examine it more minutely. On croſſing the river I was cloſe to the rock, and found a fiſſure or vein ſituated exactly in the centre, by which the ſtratum was broke and lifted up in the middle; conſequently appearing as if thrown down on each ſide, which ſufficiently accounts for the appearance.

As every circumſtance attending the ſituation of objects is intereſting to natural hiſtory, it may not be improper to give a ſhort a ſhort deſcription of the form and appearance of the caverns, which are peculiar to this ſtratum in this county; and are objects of great curioſity and admiration.

The character and form of caverns have not been noticed by any to my knowledge. If they were more accurately noticed, it might probably be the means of throwing more light on their formation. The entrances into many caverns are ſpacious, the openings are large, more particularly thoſe from the ſurface, as Peaks hole: while others are found by mining, conſequently the entrance to them is no larger than neceſſary for the purpoſes of the miners. The entrance and roofs generally aſſume an arched appearance, and though the tops of the caverns are frequently irregular, they almoſt always form a ſegment of a circle; the ſides generally riſe nearly perpendicular, while the bottoms are more

more flat. Large detached maſſes of lime-ſtone frequently lie at the bottom in rude forms: marine figures preſent themſelves in abundance, projecting in many places above an inch from the rock; chert, or hornſtone in nodules, and various forms, appears prominent in every direction. Caverns in the interior are frequently found above 200 feet high, and probably much higher, inclining to the form of an inverted cone. A prodigious variety of round or ſpherical holes occurs in the roofs, ſome two, three, four, and ſix feet diameter, and as deep; they preſerve a very correct round form: and often ſmaller ones appear in them, as if formed by art. In various places the rock forms feſtoons, and where it hangs from the roof, it frequently is extremely thin (as if worn by water) and aſſumes the appearance of drapery. The ſides and roof of caves are commonly

covered

covered with ſtalactitic matter, and ſometimes elegant ſtalactites are formed 3 or 4 feet long, and not more than one in diameter, quite tranſparent; when the infiltration of water is great, ſtalactites ſeldom appear, the rock being covered with a thick muddy marl; ſtreams of water generally occur at the bottoms, and water frequently filters down ſome part or other. In the caverns are depoſitions of ſand, earthy matter, and a variety of rounded ſtones, &c. which clearly prove, that water from a remote part has found a ſubterraneous courſe into theſe caverns, and probably was the principal agent at ſome period of their formation. Openings or ſwallows frequently occur of conſiderable depth, ſome are diſcloſed from the ſurface, as Eldon hole, others are found in mining: they are generally uneven at the ſurface, and the ſides are commonly perpendicular; they appear to be a part

of

of the ſtratum ſunk, and to have filled ſome cavern below.

The limeſtone ſtratum is found frequently divided by the toadſtone, which I ſhall now notice. It forms the ſurface in various parts of the county, beginning in the neighbourhood of Matlock, and dividing the limeſtone for a conſiderable diſtance near Buxton, and particularly at Wormhill; in that neighbourhood, it is of conſiderable extent, uneven, and rocky, but by no means ſo much ſo as the preceding ſtratum. This ſubſtance is ſingularly acted upon by the atmoſphere, and puts on ſuch a variety of appearances and difference of characters, as to render it difficult to know it in its various ſtages of decompoſition; in ſome places it appears like baſalt, or rather what is called whinſtone; abounding with hornblende, and in it are found jaſper and calcedony. At a ſmall diſtance not exceeding 20 yards

yards, it migrates into a variety of amygdaloid, ſome dark green and hard, others ochre yellow, with globules of green earth; and is as ſoft as clay.

It is very probable this ſubſtance was at ſome early period equally hard; but from being ſo differently expoſed to the action of the atmoſphere, is in ſome places covered with vegetable earth, moſs, &c. and in other places it may receive the filtrated water from the limeſtone ſtratum, which perhaps may, in ſome degree, be the cauſe of its various appearances.*

It is not conſidered as a ſtratum that admits of water filtrating through it, though for a ſmall depth it is penetrated by it; ſprings

* This ſtratum is conſidered by the miners as very uneven; it by no means ſo frequently divides the limeſtone and veins of ore as is imagined; a number of mines from 50 to 80 fathom deep are ſituated in the limeſtone, where the toadſtone has not been met with. In fact, it may be ſtated ſeldom to occur inſtead of generally.

often

often appear on its ſurface. To give an exact account of the variety of appearances this ſubſtance takes in the ſame ſtratum, would be too extenſive for a work like this, neither is it to be ſuppoſed every place is mentioned, where it and the preceding ſtrata, make their appearance. This is intended to give a conciſe deſcription of thoſe only that have been under my obſervation, more particularly in the neighbourhood of Caſtleton.

SECTION XII.

Some account of the Mines north of Derbyſhire.

HAVING thus briefly deſcribed the mines and mineral ſubſtances of this county, perhaps a ſhort account of the mines further north may prove acceptable, and ſerve as a guide to thoſe who wiſh to viſit mineral countries; my wiſh being to impart ſuch information, as may aſſiſt the progreſs of mineralogy.

The coal mines at Wiggan, about 16 miles north of Mancheſter, are worthy of notice; for here is found the noted *kennel* or candle coal.

Near

Near Chorley, 25 miles north of Wiggan, are lead mines, not now worked, belonging to Sir F. Standiſh. Theſe mines, which produced the witherite or aerated barytes, were ſunk in the grit or ſandſtone; they ceaſed to be worked about 15 years ago, and are now filled up by the earth running in, or are full of water. It would be fortunate if their produce could be rendered more uſeful; and I have been informed, that the proprietor offers liberal terms to adventurers.

To the weſt of Lancaſter, is Ulverſton, remarkable for iron mines of rich hæmatites. One perpendicular vein of ore is thirty yards wide, in limeſtone; large nodules, ſome even weighing four cwt. of a kidney form, metallic luſtre, and ſtellated fracture, are found in the looſe ore. This iron is peculiarly ductile, when it is melted with charcoal; and it is uſed for making wire. Part of the ore is tranſported to Bunawe, in the highlands of Scotland, where a company has eſtabliſhed a foundery, fed with charcoal, ſmall wood being plentiful near this diſtant eſtabliſhment.

From Ulverſton to Conniſton; near which there is a copper mine of little conſequence, and ſome ſmall mines of lead. The mountains are chiefly of the blue argillaceous ſhiſtus.

Hence to Borrowdale, in which is the remarkable mountain containing what is underſtood by black lead, or as erroneouſly termed by the miners *black wad*; an important article of commerce, and which may be ſaid to ſupply all Europe with the black lead pencils. This mine is ſituated on the ſummit of a granite mountain, high, and difficult of acceſs; lower in the mountain is driven a ſhort level.

The *wad* lies in what the miners call *locks* or ſmall cavities; and forms an irre-

gular

gular kind of pipe vein, attended by ochreous matter, calcareous ſpar, quartz, and more generally by a greeniſh, ſcaly, ſoft micaceous earth, appearing a little like a ſpecies of ſerpentine. This mine is only worked once in two or three years; a method which produces a ſufficient ſupply for that ſpace of time. The mine is carefully watched, and a houſe is built over the entrance. I believe that it is felony even to take ſpecimens from the hillock. Plumbago, or black lead, contains about ninety per cent. of carbon, combined with one-tenth or one-eighth part of metallic iron.

From Borrowdale to Kendal, eight miles, a very mountainous country. Here is a very fine variety of green granite; an uncommon ſubſtance, alſo found in the county of Galway, in Ireland. It is ſometimes uſed

 for

for flooring: but the rocks are generally a kind of blue argillaceous ſhiſtus, which on expoſure to the atmoſphere, decompoſes in thin laminæ. The geologiſt will find many intereſting varieties of rock in this neighbourhood. At the bottom of Derwent water, the lake ſo called, are a few veins of quartz in the ſhiſtus mountains, containing lead ore; but the produce is ſmall and of no moment. In the vicinity is a fine ſpring of ſalt water, not hitherto noticed. It will be worth while to viſit Croſsthwait's muſeum, where the traveller will meet with intereſting ſubjects, and procure information with civility and attention.

From Keſwick to Caldbeck are ſome mines which were formerly worked, but there is nothing intereſting to ſtop the traveller. Hence to Carliſle; near which

place

place is a vein of antimony, and a mine worked for that ſubſtance, which I did not ſee.

SECTION XIII.

Account of ſome Mines in Scotland.

FROM Carliſle I paſſed to Moffat, noted for its baths and mineral ſprings. Hence to Elvanfoot bridge, and to Lead Hills, a mineral country belonging to Lord Hopton. The mountains are compoſed of what is improperly called whinſtone, for it ſeems rather a ſiliceous ſhiſtus, or perhaps a kind of baſalt; it is of a very dark brown colour and very cloſe texture, ſmell earthy when breathed on, fracture ſplintery, and irregular; is very hard, gives fire with ſteel, and is very ponderous.

The

The veins of lead ore are in general large, and extremely rich. The susanna vein is the admiration of travellers, being a great rake vein, which in some places has continued for a considerable way, 14 feet wide of solid ore. It is now full three feet wide; and an amazing quantity is before the miners. This mine is about 100 fathoms deep, with a fire engine, not now employed, a sufficient quantity of water having lately been procured to work the water engines, so as to keep the bottom dry. There are many other veins, and in general rich; in some parts are veins of a granitic varieties of a fresh red colour, containing cubic pyrites: they appear in all directions when they approach the veins of lead ore; they do dot intersect or divide them, but continue their direction, forming the gangart on each side, they are from a few feet to 35 yards wide, intersect-

ing

ing the whinſtone; a ſingular circumſtance which alſo occurs in Brown's vein, and the mines at Wanlock Head, &c.

The veins produce oxyd of copper, mountain cork, and great variety of carbonates, phoſphates, and molybdates of lead, with quartz and calcareous ſpar; and ſometimes barytes.

At Wanlock Head, diſtant two miles to the ſouth, are ſome rich mines belonging to the Duke of Queensberry; the ſubſtances being much the ſame as at Lead Hills.

Near the latter mines native gold is procured in grains, by a ſimple, but intereſting procedure. A perſon who underſtands the buſineſs, (and little art is required,) takes about 20 or 30 pounds of earthy matter, as near the ſolid rock as poſſible, and putting it in a trough, he proceeds to a rivulet. Filling the trough with water, he turns the earth a little with a ſmall ſhovel, and

places

places it in the current, which carries off the light earth; and the ftones are thrown out. This procefs is repeated till little be left; which is turned again and again, in the running ftream, till it diminifhes to a few ounces, chiefly of lead ore and other heavy particles; among which, a few grains of gold are almoft always difcovered, their gravity precipitating them to the bottom.

On enquiring if it would be productive to conduct thefe operations on a large fcale, I was anfwered in the negative. The fmallnefs of the grains indeed afforded fufficient information; yet the certainty of finding gold, more or lefs, is ftrikingly fingular. Pieces have been found above an ounce in weight, and I am informed, that Lord Hopton has a piece yet heavier in his poffeffion. A company of Germans once worked here for gold; and the ma-

nager

nager is ſaid to have made a large fortune; but this may be a popular ſtory. It is certain, that gold medals were ſtruck from the produce, when Charles I. was crowned at Edinburgh.

The miners at Lead Hills are men of good morals, and excellent members of ſociety. They have a pretty large circulating library, which they founded for their own amuſement and inſtruction. The agents are men of ſcience, who exert themſelves to promote induſtry and happineſs.

From Lead Hills I paſſed 50 miles to Glaſgow, where there is plenty of coal found in the neighbourhood. Hence to Tyndrum, where there were conſiderable lead mines that produced vaſt quantities of lead ore, with a variety of other mineral ſubſtances, and uſed to employ a great many people. The veins are in a mountain of granite;

granite; but the operations have ceafed for fome time.

From Tyndrum I went to Strontian, about 70 miles, over the black mountains, the Devil's ftaircafe, and the grand *Glen Co*,* where there are many curious porphyries, and granites, and their varieties.

Near Ballyhulifh is the largeft and beft flate quarry I ever faw, in a mountain of micaceous fhiftus, with a feam of limeftone in the vicinity. Thefe flates form a confiderable article of export, the limeftone is burnt; but coal is fo dear that the advantage cannot be general.

Strontian is fituated in a fine valley, near the bottom of Loch Sunart. The mines are in the mountains, at the upper end of the vale, the rock being red granite; but the neighbourhood contains many va-

* *Glen Co*, near Loch Lung, contains micaceous fhiftus.

rieties of this ſtone, ſome very full of mica. The veins are what the miners call *rake*, and of conſiderable extent. Some of them have been worked about 100 fathoms ſunk in the vein, without any perpendicular ſhaft. The companies that have hitherto worked theſe mines have always been unfortunate; and they are now full of water. Beſides lead ore, was produced *ſtrontian*, a rare and new variety of earth; with calcareous cryſtallizations, zeolite, ſtaurolites, &c. but none of theſe ſubſtances are now to be found. I was there in July 1800, and could not procure one ounce of the ſtrontianite; and was informed, that two or three people had, for more than a year, been employed in picking it up, wherever it could be found.

Ben Riſſabel, in the ſtatiſtic account of Scotland, Vol. XX. p. 289, it is called

Ben

Ben Reiſipoll, and the height is ſaid to be 887 yards. A very high mountain of white granite, is about ſix miles to the weſt of Strontian; the ſummit preſents a micaceous vein, containing large garnets.

I returned by Collendon to Stirling. In the neighbourhood of the latter place the Ochill Hills, eſpecially near Alva, contain veins that produce ſilver, copper, cobalt, lead, &c. The mines are not now worked; but the naturaliſt who takes the trouble of picking the hillocks, and ſearching the mountain, will meet with ſome recompence for his labour. At Edinburgh is Weir's muſeum, where I hope to ſee mineralogy more noticed; the largeſt departments are the birds and the fiſh. Arthur's ſeat, and Saliſbury Craigs, are of baſalt, ſometimes forming rude irregular pillars.

The whinſtone, or baſaltic mixture of quartz and hornblende, is uppermoſt; and under it is a ſeam of jaſper, in ſome places curiouſly ſpotted with iron. Here alſo occur grit and limeſtone; with quartz and calcareous ſpar. When the whinſtone is in decompoſition, it appears like a lava that has been long expoſed to the weather, and much reſembles the Derbyſhire toadſtone. The Pentland hills produce *petunſe*, or the decayed felſpar uſed in making porcelain; and, according to ſome, adularia. Here are alſo conſiderable maſſes of argillaceous porphyry.

From Edinburgh I returned by Carliſle to Alſton, where there are ſeveral mines belonging to Greenwich Hoſpital; from the truſtees of which, they are rented by the Quaker company. They are chiefly in limeſtone; and are rake veins, producing

producing a great quantity of lead ore, with blende, calcareous ſpar, &c.

Garrigill gate, Tyne head, and Nent head, are conſiderable mineral countries. Coal Clough, a great mine, is worked in lime-ſtone, and in the grit ſometimes called *hazel*; and alſo, which is ſingular, through a thin ſeam of coal. It produces large quantities of lead ore, fluor, &c. There is another conſiderable mine of the ſame ſubſtances at Allons Head; both of theſe mines, and many others, being the property of the worthy Colonel Beaumont. The mines in this quarter are conducted on the moſt ſcientific principles, which their rich produce can indeed well afford; and the agents are men of ſkill, and well verſed in mechanics. On proceeding ſouthward, a variety of mines appear in Weredale and Teeſdale, about Middleham and Ark-

 endale

endale in Yorkſhire, and at Kettlewell in the vicinity; nor muſt I omit to mention thoſe at Graſſington and Paitley bridge.

SECTION XIV.

Tour from Glaſgow to Staffa.

SHOULD the traveller wiſh to viſit the celebrated iſle of Stáffa, one of the weſtern Hebrides, the following ſhort account may perhaps prove intereſting. I ſhall not attempt to delineate the vaſt extent of mountainous country I travelled over: that would require a work of time, and could not be accompliſhed in few words.

From Glaſgow I determined to make a pedeſtrian tour to Staffa; accordingly I ſet out for Dunbarton about 15 miles diſtance, good road; the neighbouring rocks appear

 a dark

a dark baſalt ; from Dunbarton I proceeded by Loch Lomond and Ben Lomond ; excellent road and fine rock ſcenery to Luſs, and from thence to *Arrocher*, 22 miles. On the ſide of Loch Long a good inn; where I arrived about 6 o'clock in the evening. The day having been rainy, and the evening commencing with violent gales of wind, accompanied with heavy rain, cauſed me to determine to ſtay here until morning; having picked up a few fragments of rock ſpecimens and carried them in my pocket, which I now began to unload and examine their merits, in order to throw away thoſe which was leaſt intereſting. They chiefly conſiſted of baſalts, argillaceous ſhiſtus,granites,and their varieties. The argillaceous ſhiſtus I found frequently in veins in the baſalt ; it appeared in various directions, and falling in laminæ by decompoſition; in it were many quart-

zoſe

zoſe veins, and patches of fine blue ſlate. A beautiful red granite, in which the felſpar was of a high fleſh colour, was amongſt them; the felſpar compoſed the greateſt part of the ſpecimen; in one or two pieces of granite, the mica compoſed three parts out of four of the whole.

I ſhall not attempt to amuſe the reader by telling how I ſpent the evening, what I had to eat, &c. ſuffice it to ſay, the inn is a good one, and the hoſteſs did credit to her houſe.

In the morning I aroſe at 5 o'clock, and immediately ſtarted for Cairndow, 12 miles diſtant. Having walked round the head of Loch Long, my mind was ſtruck with awe at the approach of Glen Cro; the roads are excellent, and the traveller is amuſed almoſt every mile by an inſcription engraved on the mile ſtone; by whom

 they

they were made, when they were erect-ed, &c.

Glen Cro is a very deep ravine, ſurround-ed with vaſt mountains extremely rugged. The morning was exceedingly rainy with a violent high wind; a rivulet runs thrugh the Glen, which in wet weather, muſt be conſiderably increaſed from its rapid fall. Immenſe caſcades are formed in heavy rains, precipitating large pieces of rock to the bottom, which lie in all directions.

The Glen is ſo much on the aſcent, that there is not an acre of even ground in the diſtance of ſeveral miles. At the top is placed a ſtone, on which is en-graved, "reſt and be thankful:" certainly thankful for a good road, but the aſcent is not ſo great to men accuſtomed to moun-tainous countries, as to be ſo much ſin-gularized. The rocks are chiefly a mica-

ceous

ceous ſhiſtus, frequently containing large veins of quartz.

Onward I proceeded this unfavourable morning over a variety of mountains and glens; whoſe beauty was in moſt parts hid by the inclemency of the weather, until I arrived at Carndow, on the banks of Loch Fine, a delightful ſituation; I dried my clothes, breakfaſted, and received every comfort a good inn could afford a wet traveller.

Being refreſhed, I ſtarted at 11 o'clock for Inverary, diſtant about 12 miles, the weather ſtill unfavourable. In mountainous countries, when it begins to rain, it frequently continues a long time; but in walking a few miles the atmoſphere is commonly leſs agitated. If the traveller ſtops every day, he meets with bad weather, it will be long ere he accompliſhes his journey in this country. Walking by the head of Loch Fine, I picked

picked up many varieties of granite, which added confiderably to my weight, and were very inconvenient to carry. I examined many fiffures in Bafaltic rocks as I paffed, and found the fides corroded, and very full of holes. Arrived at Inverary, refrefhed, and proceeded to Port Sonnochin, 12 miles diftant. The Duke of Argyles' feat is delightfully fituated, and happily adapted to its fcite and furrounding objects; the interior is elegant and magnificent. The caftle is built of a fingular kind of ftone, I fcarcely know by what name to call it; it appears a fpecies of argillaceous pot ftone; the magnefian earth feems predominant from its foap like feel. I was informed that it is found on the oppofite fide of the Loch. I proceeded to Port Sonnochin, over immenfe mountains immerfed in the clouds; and was furprifed to fee fuch quantities of rounded pieces of granite,

fome

ſome of amazing ſize, detached on the ſummits.

Port Sonnochin is on the banks of Loch Awe, a freſh water lake well ſtored with ſalmon ; I was ferried over about half a mile, and proceeded to Bunawe, diſtant about 10 miles. Having croſſed the lake, the country opens, and the ſcenery is leſs wild; ſhiſtus and granite ſeem to compoſe the general range of mountains, which are here covered with ſmall wood. At 9 o'clock arrived at a tolerable good inn, pleaſantly ſituated, where I ſtaid all night.

In the morning I was gratiſied with fine weather; ſet out for Oban, diſtance about 12 miles, a moſt excellent road and pleaſant walk, by Loch Etive to Dunſtaffnage.

Here the glens are more ſpacious and more cultivated, the mountains appear in general to be a bluiſh ſhiſtus and granite. At Bun Awe is a ſmall iron work, ſituated

here

here for the convenience of ſmall wood, which is converted into charcoal. On approaching Oban I met with a ſingular rock, preſenting its bare perpendicular front on the road ſide, and of conſiderable extent; it appeared to be compoſed of an aſſemblage of large ſtones in all ſhapes and directions, cemented together by an earthy kind of matter; many detached pieces lay in the Glen, and it forms the bed of the water I croſſed at the entrance of Oban. The ſtones that compoſe this rock are chiefly granites, ſhiſtus, and quartz, rounded in a pebble-like form; ſome are very large; the mountain may be called a ſpecies of pudding ſtone, though the cementing matter ſeemed to me to be very ſoft, in ſome places, decompoſing and earthy.

This kind of pudding ſtone is in contact with ſhiſtus; in which laſt ſtratum good ſlate is found in the neighbourhood.

Near

Near Kerrara Ferry is a mountain of ſhiſtus containing veins very full of mica; on the ſummits of the rocks are many rounded pebbles, and near Oban are varieties of amygdaloids.

Next day croſſed Conol Ferry, and viſited Berigonium; near which place is a bare rock projecting on the road ſide, compoſed of the before-mentioned rounded ſtones in great variety.

Berigonium is a ſteep mountain, compoſed of quartzoſe ſhiſtus, ſituated on a large tract of low flat land, extremely boggy, from whence quantities of peats are cut for fuel.

The mountain is not large, but rather difficult of acceſs; on its ſummit is a variety of ſcoria and pumice like matter, cementing fragments of granite, quartz, &c. ſome of which are covered with enamel and other evident marks of fuſion. Theſe vitrified ſubſtances appeared to lie in maſſes in different parts. This mountain has been repreſented as Volcanic, and theſe ſpecimens (particularly thoſe that reſemble pumice) have been brought forward to ſubſtantiate the fact.

I ſhould rather ſuppoſe large fires have been made on it as ſignals to the neighbouring iſles and other purpoſes; for which its ſituation is particularly convenient. It is near the ſea, which throws up innumerable rounded ſtones, ſome of which I think moſt probably have been carried to the top of this mountain, to ſerve as a baſe or fire place.

It is alſo probable, that wood and weeds left by the tide, were gathered to burn as occaſion required, which may in ſome degree account for the vitrification encompaſſing the rounded ſtones, more eſpecially as they are only in ſmall heaps. This mountain

tain has not the ſmalleſt veſtige of any thing like volcanic origin; at one end are a number of ſtones placed without order or regularity.

From Oban I hired a boat to the iſle of Mull: and walked to Arros. In various directions I travelled about 50 miles on that iſland, and found it generally of baſalt, in ſuch various ſtages of decompoſition, as to render its character loſt in many inſtances, frequently migrating into a ſoft toadſtone. On the weſt ſide of the iſland, the cliffs adjoining the ſea have a columnar appearance.

I ſtopped at a houſe at Lagan, a tolerably good inn: they are very civil people, and the hoſteſs ſpoke Engliſh: they rent the iſle of Staffa, and accommodated me with a boat and neceſſaries for the voyage. In viſiting Staffa, I did not perceive much danger, the weſterly winds

often

often prevail and render it impoſſible for a boat to go, but in tolerable good weather (and ſurely nobody would go 10 miles to ſea in bad weather on pleaſure) a boat may approach it with ſafety.

I have viſited Staffa ſeveral times, and never met with any thing like an accident; the landing is perfectly eaſy and ſafe, when conducted by the people accuſtomed to the iſland.

Staffa is a bold high iſlet, riſing nearly perpendicular in many places; being about a mile long, and one eighth of a mile broad. It is almoſt ſurrounded by perfect baſaltic columns in different directions, and of unequal magnitude; they are in general perfectly diſtinct and detached from each other. The more earthy parts in the hollows conſiſt of a ſingular ſpecies of mandel ſtone, of a dark dirty brown colour, full of holes, many of which contain calcedony,

dony, zeolite, and olivin. Zeolite appeared very ſcarce, and I never found any deſerving the name of a good ſpecimen.

The ſummits of baſaltic priſms appear above the graſs in one part of the iſland. The boat generally lands you on baſaltic columns, which are even with the water; from which you walk on others riſing in regular ſucceſſion and ſerving the purpoſe of ſteps, until you arrive at the height of the iſland.

Near the cave of Fingal the columns are of great height, ſome perpendicular, others bending. Oppoſite there is the iſland of Booſhala, which is compleatly formed of columns in all directions; but leſs than thoſe of Staffa, from which it is ſeparated by a narrow ſound, very deep, though not exceeding 10 yards wide.

The ſea almoſt continually beating againſt the weſtern end of the iſle, may probably

have

have formed the cave which is ſituated there; it may be entered with a boat in fine weather, the water in it is deep, and a great ſurf runs in high winds. The approach and entrance to the cave are by walking on baſaltic columns, that alſo form a path to the end, which may be 60 or 70 yards.

Its breadth at the entrance about 12 or 15 yards, its height about 20 yards, depth of the water in the cave, from 10 to 15 feet.

The conſtant humidity of the cavern, cauſes the tops of the columns which form the path, to be extremely ſlippery; they are alſo at unequal diſtances, and unequal in height, ſome being a foot or more higher than others, and the width being only the diameter of a column which renders this not one of ſafeſt roads for a traveller, as one ſlip would plunge him into 10 or 15 feet water, with the additional danger of a violent ſurf

that

would render ſwimming uſeleſs. I would adviſe the viſitor, whoſe curioſity may lead him to the far end, to take off his boots, by which he will have the uſe of his feet better, and be leſs liable to ſlip. This iſland, though bare of ſoil, produces good graſs, and is much eſteemed for paſture; ſometimes 20 or 30 head of cattle are feeding on it.

One family reſides here to take care of them during the ſummer. In the neighbourhood are ſeveral other intereſting iſlands, as Iona the luminary of the Eaſt many centuries ago; here lie more than forty Scottiſh, Iriſh, Daniſh, and Norwegian Kings or Chiefs; here are alſo the remains of many monuments and of a cathedral.

Dutchman's Cap is a ſingular iſland, alſo Ulva and others adjoining to Mull.

SECTION XV.

Salt Mine of Northwich.

ON proceeding from Kutsford to Northwich, the chief appearance is that of grit ſtone. with ſome nodules of granite and clay.

A ſmall diſtance from Northwich I was lowered down into the ſalt work, about 110 yards, by a ſteam engine. The ſhafts are lined with wood; and the firſt bed of ſalt is about 40 yards thick, being excavated about ten feet in height, to an extent of 50 yards.

Deſcending about 60 yards further, you arrive at the bottom, which appears like a grand

grand amphitheatre, about 120 yards long, 100 broad, and 15 feet high. Triangular pillars of a vaſt ſize ſupport the roof, and every part being perfectly dry and clean, the effect is very pleaſing.

They get the ſalt with long chiſſels, and by the blaſt, working from the ſolid rock at the ſides.

The ſtrata are extremely irregular. At the ſurface is a ſtrong red clay, then a ſhiſtoſe ſubſtance called *plate*, which is followed by a little layer of coarſe grit; then a blue clay with gypſum, to which ſucceeds a ſubſtance called *metal*, being an indurated clay, in which are veins of rock ſalt. This clay is of great thickneſs, and is followed by a bed of rock ſalt, workable for perhaps 20 feet. Next is a ſtratum of clay, mixed with ſalt, about 30 yards thick, more or leſs impregnated, until it becomes a hard compact rock ſalt, which is generally of a reddiſh brown, and ſometimes in thin layers perfectly tranſparent. It is worked at an expence of about two ſhillings a ton.

To draw up the rock ſalt, they have fire engines, equal in power to three hories. It is chiefly exported under a heavy duty; and in the town are fifty or ſixty officers of exciſe.

Common ſalt is procured by diſſolving the rock ſalt, and if a fine ſort be required, a ſtrong fire is made under the brine, and the ſalt forming at the top, precipitates to the bottom. If ſtronger ſalt be wanted, as for fiſh, &c. the heat is adminiſtered ſlowly, when the ſalt cryſtallizes in large cubes.

The whole ſurrounding country is clay, and very coarſe grit; and wherever a ſhaft is ſunk in the neighbourhood, there is a certainty of finding ſalt.

The

The grit ſtone prevails as far as the town of Flint in the county ſo called; and as no coal is found near Northwich, it is preſumable that the ſalt is above it.

In the county of Flint, and near the capital of that name, are plenty of coal mines at various depths.

Towards Conway there is abundance of limeſtone, and at Newmarket I obſerved ſome heaps of green calamine.

Llandidno is ſituated on the top of a hill, and is a poor village. The copper mine is in limeſtone, and ſeems to occupy a large ſpace; the ſurface has ſunk ſeveral yards, the bottom being ſoft, and containing caverns, with calcareous ſpar and copper.

This mine is ſaid to have been worked by the Romans, and ſeems to lie in bunches. There are ſome curious calcareous ſpars; and the limeſtone is very full of chert. Fine malactites, of the velvet kind, have been found here; and the ſandy ſort of copper ore.

From Conway to Beaumaris are chiefly blue ſhiſtus and granite; and near Bangor are good quarries of ſlate.

SECTION XVI.

The Paris Mine.

BEING now arrived in the iſland of Angleſea, I was anxious to ſee the celebrated mine in the mountain of Paris. The ſmelting works firſt attracted my obſervation, being ſuperior to any I had ſeen, and containing 20 furnaces in a very extenſive building.

The mine is on the top of a mountain of blue or perhaps quartoze ſhiſtus, or perhaps ſome might call it a quartoze ſhiſtus with ſerpentine; ranging from eaſt to weſt, about 500 yards in length, while the breadth is about 100 yards, and the depth nearly as much. The bottom is very irregular, maſſes of *rider* or vein ſtone interfering; while the richer copper ore runs into holes, and crevices, in ſtrange and various directions. The ſhiſtus lies in irregular ſtrata, and is covered with a bed of gravelly heterogenous matter, full of chert.

Copper ore is got for about two ſhillings a ton, and is laid in heaps of five or ſix hundred tons; in the ſides of which ovens or fires are placed, and the ſulphur in the ore ſoon taking fire, it continues roaſting for ſix or nine months, and is then forwarded for ſmelting. The produce of the mine is very poor, about ſeven and a half per cent. ſometimes more, and ſometimes ſo little as five per cent. Patches appear of fine cubic mundic.

Great quantities of ſulphur are made. In the ſubliming houſes the ore is covered

vered with earth, and brick tunnels are formed on the top or ſides, to receive the ſulphur.

On examining this immenſe mine, it does not appear like a vein. About 500 yards to the eaſt, the Corniſh Company have ſunk a ſhaft of 40 fathom, but have only found ſmall particles, and ſtrings of ore. They have alſo driven north and ſouth; but have not met with any vein.

In ſome places the ore riſes within eight inches of the ſurface; and immediately above is a red cruſt, which has every appearance of vitrified ſcoriæ, has marks of fuſion, being cellular, glazed, and like a feruginous pumice. It is ſometimes a little irredifcent, has a ſtyptic vitriolated earthy ſmell; and probably contains vitriolic acid, with iron and ſometimes lead.

The area of ground, containing the whole works, is at leaſt a ſquare mile in compaſs.

The

The water, which is pumped from the works, is ſtrongly impregnated with copper, and is received in dams and reſervoirs conſtructed for the purpoſe, and in vats like the pits uſed by tanners. In theſe are placed plates of caſt iron, on which the copper is precipitated, which is of the beſt quality. Of theſe pits there are ſome hundreds; and in the vicinity are ovens to dry the copper.

After the ſulphur is reſined, it is melted in iron veſſels over a ſlow fire; and when liquid is caſt into moulds, perhaps earthen boilers would improve the colour, which is far from good.

The Paris mine is worked at a conſiderable expence; and with the Mona mine, employs about 1,200 hands above ground. The greater number are occupied in ſorting, dreſſing, waſhing, &c. while few are employed under ground, the deepeſt ſhaft being about 40 yards.

The

The copper ore is extremely full of ſul-phur and iron. About four hours are employed in reducing it to a regulus. The metallic ſubſtance running chiefly into the firſt pig, which is richer than the reſt, the regulus ſinking from its ſuperior weight. The ſecond is tolerable; but the others are thrown aſide, and form an article of ſale for mending roads. Formerly conical kilns were uſed, but they were found too expenſive.

The ſmelting houſe is about 80 yards in length, by 18 in breadth and about 6 yards high.

The harbour is ſafe and convenient; but a bar ſeems wanted to break the ſurf.

The ſurrounding mountains appear to be all formed of ſiliceous ſhiſtus; with a kind of ſerpentine, which cleaves and decompoſes.

The

The chief minerals I obſerved here are as follows.

A kind of lead ore reſembling clay.

Sandy lead ore.

Yellow copper ore.

Black ditto.

Irideſcent ditto.

Varieties of mundic.

Copper formed by cementatio

Native vitriol.

Ochres.

Lead ore ſulphuric.

Steel grained galena.

A ſhiſtus ſerpentine, with aſbeſtus.

Green ſemi-tranſparent aſbeſtus.

A quartzy ſhiſtus, full of veins of white quartz.

Many varieties of chert, with veins of copper in them, forming a part of the riders in the mine.

Allum

Allum is alſo prepared here, from ſhiſtus laid on the roaſting copper ore, where it abſorbs the vitriolic acid. The pans uſed are of lead.

SECTION XVII.

Some other Obſervations in Wales.

RETURNING by Conway, I inſpected the mines in that neighbourhood, but found nothing worthy of notice. Their matrix is chiefly chert, a ſpecies of black hornſtone; and there is a black flinty ſtone with white quartz.

The vale of Conway is extremely picturesque, and full of rich ſcenery, the rock and waterfalls conducing greatly to the effect.

Near Llanruſt are ſome large veins of quartz, which are worked for the porcelain

lain mill, near Bangor; and I ſhould have mentioned, that near Conway are ſome maſſes of chert, in a black ſiliceous mountain, which are found nearly as well adapted for mill ſtones as the French *burr:* and a lady obtained the premium of 100 guineas, from the ſociety of Arts, for the diſcovery.

The mountains near Llanruſt are of granites, porphyry, and ſhiſtus, with a great deal of black chert, in which are veins of lead ore, quartz, &c. but there are no lead mines of any conſequence.

I re-viſited Llandidno, and went down the mine. There are veins of copper ore, which ſometimes preſent that ſubſtance, and at others maſſes of limeſtone. Here is a vaſt *lum,* filled with looſe earthy matter, diſlocating the ſtrata.

SECTION

SECTION XVIII.

An account of Mr. Williams's Book, called the Mineral Kingdom.

Edinburgh 1789, 2 Vols. 8vo.

Vol. I. 450, p. p. Vol. II. 531.

THIS work is written in ſo ſingularly prolix and confuſed a manner, that an analyſis of it may be uſeful to thoſe who poſſeſs the work, and to thoſe who do not, may afford an inſtructive account of its contents.

The title page expreſſes that it is divided into three parts:

"PART I. Of the ſtrata of coal, and the concomitant ſtrata."

"PART

PART II. Mineral veins and other beds, and repositories of the precious and useful metals.

PART III. Of the prevailing strata, and of the principal, and most interesting phenomena upon and within the surface of our globe.

Of these Part II. which occupies about a third part of the first volume, is the most curious and interesting; but like the rest, disgusts by its tedious prolixity, and want of scientific arrangement.

At the end of the second volume is a table of contents which ought to have been prefixed to the work. There are neither chapters nor sections, so that the reader is bewildered in a vast mass of matter; and this neglect has led the author into many repetitions.

In the preface, the author explains the importance of our coal and other mines to the manufactures and commerce of the nation;

tion; and the consequent importance of acquiring mineralogical knowledge. He shews that many manufactures of the greatest consequence depend entirely on mineralogy, and in page xxiii. begins a confutation of the Huttonian system, which continues to the end of the preface, p. lxii. A chief end of this analysis, is to select the real facts and observations, which in this work, are buried in a mass of idle declamation. Such is that, p. xlix. that the rich vein of lead at Llangunog in Montgomeryshire; which was five yards wide of solid ore, was found suddenly cut off by a deep bed of shistus. This preface concludes with a prayer, the author being a very good christian.

PART I.

The ſtrata of Coal and its Attendants.

THE author firſt inveſtigates the regularity and inclination of coal ſtrata. The incumbent ſubſtance is called the roof, while that below the coal is termed the pavement, which generally proceeds in a manner ſurpriſingly regular, except when interrupted by *troubles*. Theſe conſiſt of ſlips, dykes, gaſhes, and other circumſtances. The moſt common is the ſlip of the Engliſh colliers, the dyke of the Scotiſh, in which the ſtrata are broken aſunder, and thrown up or down on the other ſide of the ſlip. Parallel ſlips cut the ſtrata in their inclination, which is generally from north eaſt to ſouth weſt. Oblique ſlips paſs acroſs the ſlope. Direct ſlips cut the ſtrata right acroſs. All theſe ſlips would be called by naturaliſts, perpendicular fiſſures, and by miners, rake veins. A hitch is a ſmall ſlip only affecting part of the coal; his other remarks on this ſubject, muſt be intereſting to coal proprietors.

The next interruption is that properly called the dyke, conſiſting of hard ſtone, commonly called whinſtone, of ſofter ſtone, clay, gravel, or ſand; the name ſeems to ariſe from its reſemblance to a wall in the north, called a dyke. The ſofter and looſer ſorts produce much water, which renders them difficult to be penetrated. The *gaſh* or *gall* is very frequent, being a crack or chaſm in the coal, and its concomitant ſtrata; and when wide, it is ſometimes full of

looſe

loofe matter, fo as to partake of the nature of the dyke.

The fhake affects the coal and ftrata, fo as to break them and throw them into confufion, and when large, the beft expedient is to fink a fhaft upon the other fide.

Mr. Williams then proceeds to give fome account of the coal ftrata to the eaft of Edinburgh; particularly thofe near the Temple Mill, which prefent four feams of coal. The author then gives fome account of Arthur's Seat, and Salifbury Craigs, &c. and of the great regularity of ftrata on the coaft of Caithnefs. The author thinks, that moft caverns, have been mineral veins.

Among the regular ftrata, Mr. W. claffes coal; argillite and mountain rock among the irregular mountain limeftone, which Mr. Kirwan fuppofes to be primitive limeftone. Among fubftances feldom ftratified, he men-

 tions

tions granite, of which a mountain may appear as compact as a ſmall piece.

He then returns to coal, and mentions many inſtructive particulars which need not here be repeated. The roofs of coal or incumbent ſtrata, he divides into baſalt, as at Hill Houſe, a mile ſouth of Linlithgow limeſtone, poſt ſtone, or ſandſtone, dogger-band or ſtrata, in balls of iron ſtone, blaes, or black ſhiſtus, alſo form a common roof; but his further remarks on this ſubject become ſcarcely intelligible from the imperfection of his mineralogical knowledge and vocabulary. The extent of the adit at Kilmarton, which interſects above 60 beds of coal, with different intervening ſtrata, is particularly diſplayed.

Our author then enlarges on the declination of the coal ſtrata, and the various accidents to which they are ſubject, particularly their running ſometimes like waves

but

but many of his remarks are very local and minute.

He then proceeds to examine the extent of coal fields which he finds do not paſs through mountains, but on the contrary, ſometimes terminate at ſome diſtance from them. The coal field to the S. E. of Edinburgh, extends about ſixteen miles from Duddingſton to New Hall, where it terminates at the bridge of Carlops, where the river Eſk leaves the Pentland Hills; and the ſeams of coal, inſtead of paſſing under the hills, baſſet or riſe up in great confuſion. He concludes, p. 141, that the coal fields do not ſtretch under the mountains, but are patches of different dimenſions like fields of corn or graſs. That of Mid Lothian is about fifteen miles in every direction, the fourth forming the northern boundary.

After ſeveral repetitions which might have been avoided by a proper diſtribution and ſubdiviſion of his ſubject; he obſerves, that hills of whin or other concomitants of coal, are not to be conſidered as interſecting the ſtrata. The coal field in Fife (he ſays,) reaches from Stirling to St. Andrews; and is in ſome places about 10 miles broad. He then explains how coal may be exhauſted; and obſerves, that this ſubſtance is firſt mentioned in Engliſh records of the year 1234, and in Scottiſh 1291. Some remarks follow on the ſtate of coal at Newcaſtle, Whitehaven, and in South Wales, from which laſt, the mines of Cornwall are ſupplied. Further obſervations follow on the coal trade, and on the ſuppoſed exiſtence of large beds of coal in the iſland called Cape Breton. Nothing can exceed the prolixity of his declamations on this ſubject, which rarely preſent one ray of ſolid information.

In

In p. 207, Mr. Williams proceeds to give ſome inſtructions to landed gentlemen, on the real and fallacious appearances of coal, and p. 233, he particularly conſiders petroleum, which he ſays is often found in ſtratified limeſtone at a great diſtance from coal. He afterwards delivers his opinion, that coal conſiſts of antedeluvian timber.

He then, p. 242, &c. enumerates ſix kinds of coal. 1. Caking coal, ſuch as that of Newcaſtle, but ſcarcely known in Scotland. 2. Rough or rock coal, as that of Lothian and Shropſhire.* 3. Stone or ſplent coal of a ſlaty texture; common in Lothian Fife, Ayrſhire, and in ſome parts of England, (I believe that of Kingſwood, near Briſtol is of this kind.) 4. Cannel or parrot coal, as that of Wiggan, and alſo found near Edinburgh. 5. Culm or blind coal

* This kind ſeems always to border on primeval hills of porphyry, &c. as the firſt on limeſtone.

which

which neither emits ſmoke nor flame, but burns like charcoal. 6. Jet, which he ſays is found in England and other parts in detached and ſeparate maſſes; he then compares the different kinds of coal with the different kinds of wood, and points out p. 254, the extraordinary appearance of coal at Caſtle Lead, in the eaſt of Roſshire, where it aſſumes the form of rake veins, and as he afterwards explains in granite. He alſo found coal in the iſle of Mull; he returns to ſhew the abſurdity of ſupporting that petroleum and far leſs ochre, are any indications of coal.

PART

PART II.

Mineral veins and other Beds, and Repositories of the precious and useful Metals.

HE divides mineral veins into four kinds, 1. rake, 2. pipe, 3. flat or dilated, 4. accumulated. He obſerves, p. 271, that the vein at Strontian may be called a gaſh, and that it is in grey granite, in which he is miſtaken or inaccurate, faults too common in this work, for it is in red granite. In p. 274, he again returns to Llangunog, and afterwards gives an account of the lead mine at Daven Jaur in Cardiganſhire, and of mineral veins on the Scotiſh ſhores.

In his account of irregular rake veins, Mr. W. obſerves, that the beſt concomitants of ore are the ſpars, and vein ſtones or riders. His account of the ſpars is pretty accurate, conſidered as calcareous, as cawk or barytes, and quartz. In p. 288, he gives a curious account of lochs or the cavities in mines, and the beautiful ſpecimens of copper found at Colvend in Galloway. He then enumerates the ſoft ſubſtances found in veins, particularly that reſembling ſnuff, or the guhr of the Germans. Other circumſtances attending veins are enumerated with care, and in general this ſecond part of his work is by far the moſt preciſe and inſtructive; but cannot pretend to any praiſe of arrangement.

In p. 314, he deſcribes a beautiful ore of the Lead Hills, being a yellow effloreſcence near an inch deep, of a fibrous or columnar texture upon the blue galena.

After

After a full account of the rake or perpendicular vein, becauſe it is the moſt common, he proceeds p. 321 to conſider the pipe vein. The mining field of Ilay, he ſays, conſiſts of ſtring veins, and ſubjoins ſome account of them. The pipe vein he deſcribes p. 331, as varying from the horizontal to the declination of 45° or more. He obſerves that the lochs are open ſpaces, and more frequent in the pipe veins.

The flat or dilated vein or ſtreek, lie between ſtrata like ſeams of coal, and commonly occur in argillaceous ſtrata.

The accumulated vein, commonly reſembles a coal direct or inverted, and is uſually the richeſt of any.

Mr. W. then proceeds to explain the various ſlips, troubles, and other incidents which occur in metallic veins; and obſerves that ore is ſometimes found interwoven as if it were with the rock, yet

worth

worth working, of which he gives an instance at Cwmyſtwith in Cardiganſhire; metallic ore alſo occurs in the puddingſtone at Gourock, near Greenock, and in a ſingular ſtone near Loſſymouth, which he deſcribes as a compound of many fine ſtones of beautiful colours.* He then treats of float or ſhoad ore and indications of metal, on which ſubject he is practically inſtructive. The rachel or broken rock, called broil by the Corniſh miners, he conſiders as worthy of particular attention; but he is led into ſome repetitions concerning the ſoft mineral ſoils, and what the miners call *mother chun* or *guhr*. He then combats the opinion, that metallic veins are peculiar to mountains. Among the moſt productive ſtrata, he enumerates lime-

* From ſpecimens it is now found, that this rock conſiſts of petroſilex, with ſome galena and quartz cryſtal, ſo that Mr. W. has here indulged his imagination only.

ſtone,

ſtone, and what he calls the indurated argillaceous mountain rocks and granite; but he looks upon the ſecond as the moſt abundant, as affording the rich mines of Lead Hills, Tyndrum, and others in Scotland; thoſe of Cardiganſhire, of Yorkſhire, Weſtmoreland, and many parts of the north and ſouth of England.

He now falls into ſeveral repetitions concerning ſlips, &c. and in p. 408, deſcribes the various kinds of lead ore, afterwards proceeding to thoſe of copper and iron. Some of the miſcellaneous remarks here introduced are curious, ſuch as that p. 411 concerning the Roman works at Darenvawr, and the richneſs of the lead of Cardigan; that concerning the veins of copper near Old Wick in Caithneſs: and that found in limeſtone at Loch Kiſſern, upon the weſt coaſt of Roſshire, oppoſite to the iſle of Raſay; that concerning the copper, ſilver, lead,

lead, and cobalt found in the Ochil Hills, near the bridge of Allan, with the copper found at Curry, in Lothian; and Colvend, in Galloway. The irons he confiders as of two kinds, iron ore and iron ftone: of the latter, he defcribes a ftratum as of a reddifh brown colour, and it is alfo found in nodules in the argillaceous ftrata, which accompany coal.

PART

PART III.

Of the prevailing Strata, and of the principal and moſt intereſting phenomena upon and within the ſurface of our Globe.

THIS third part occupies the whole of the ſecond volume, and is diſcuſſed in a very prolix and declamatory manner.

The firſt topic is a general view of the prevailing rocks and ſtrata in Great Britain. He firſt mentions the regularly ſtratified mountain rock, as whin and argillite, among the latter, the fine blue ſlate of Eiſdale and Ballachyliſh, in the iſlands of Scotland, that of Stobo in Tweed-dale, and the purple ſlate near Tombay, above Callender. After deſcribing ſhiſtus, he proceeds to the granatic rocks, on which ſubject he diſplays little knowledge. The peaſy whin is found in Galway, conſiſting of black and white grains of the ſize of ſmall peas. He returns to the rock at Loſſymouth, and then proceeds to ſpeak of limeſtone. The mountain kind he has ſeen in the iſlands reſembling Parian marbles, " and ſome of it compoſed of fine glittering ſpangles as large as the ſcales of fiſhes." The aſh-coloured mountain limeſtone with ſmall grains he obſerved in the iſle of Ilay, and in the country of Aſſynt, to aſſume the exterior appearance of ſharp jaggs about a foot long. He then deſcribes the white ſtatuary marble of Aſſynt, and ſome other kinds. He miſtakes the ſerpentine of Portſoy for Jaſper, and a hill

of

of quartz, near Rothes, for agate.* The granite of Ben Nevis he here calls porphyry, but he adds, that about three quarters of the way up this mountain, upon the N.W. ſide, there is found a porphyry of a greeniſh colour, with a tinge of browniſh red ſpotted with white angular ſpecks. He then returns to granite, and afterwards mentions baſalt. Next are marl and chalk, and micaceous ſhiſtus; then a more ample account of baſalts.

He proceeds p. 49, to brecica or pudding-ſtone, and p. 52, to the mountains of quartz in Roſs and Inverneſshire. Next are the ſtrata of ſandſtone, particular thoſe of Caithneſs.

He then enters the wide field of the ſtratification of this globe, and mentions p. 63, the prodigious maſs of granite which

* From the ſpecimens it is a white quartz, in ſome parts cryſtallized and tinged red with iron.

 compoſes

composes Ben Nevis, which he seems to describe as being four miles in length. It is unnecessary to follow him through the mazes of theory; suffice it to remark, that he supposes veins and faults to be fissures occasioned by heat, and afterwards filled by depositions from water. He examines at some length the system of Buffon, which he considers as impious and chimerical. He afterwards investigates the structure of mountains, and points out a valuable mill stone rock near Loch Broom. One of his most singular remarks p. 152, &c. relates to the pudding rock, which he traces in Sutherland, Ross, and Inverness, &c. in the north of Scotland, and from Monteath to Stonehaven in the south; and in p. 156, that it is also found to the west of Thurso in Caithness, and the vitrified forts, as he says only occur upon this kind of rock, the account of which he amplifies p. 158, as ſtretching along the S. E. ſide of the Grampian Hills, by Kinfauns in Perthſhire, and into Dunbartonſhire, croſſing the Clyde to Ayrſhire, where it finally enters the eaſtuary of that river. Some alſo appears in the neighbourhood of Dumfries, and it ſeems palpably to have been waſhed down from the higheſt mountains.

In the confuſion of his arrangement, he next deſcribes talc and mica, and in his account of quartz and felſpar, he blends and confounds thoſe different ſubſtances. He mentions p. 175, a ſingular amethyſtine ſand on the river Aldgrant, in the eaſt of Roſshire,* but this is probably as imaginary as his rocks at Loſſymouth and Rothes. Of ſhill or ſhorl he alſo gives an imperfect account.

He afterwards particularly examines the

* He calls it the river Allgrade, and ſays it runs into Moray Firth, *inſtead* of the Firth of Cromarty.

ſtrata of coal, and on this ſubject falls into many repetitions. His theory of antedeluvian tides correſponds in ſome degree with Mr. Kirwan's geological eſſays; but his idea of antediluvian ſtrata, conſiſting of uniform mica, uniform quartz, uniform diamond, &c. deſerves little attention, and his theory is deſervedly forgotten among many others. The population and natural hiſtory of America, form a long and tedious digreſſion.

At p. 319, we find what he calls *tracts* on ſeveral ſubjects relating to the *mineral kingdom.* The firſt of theſe *tracts* is on volcanoes, and here again we find many repetitions concerning veins, &c. nor does this theoretical eſſay throw any ſtriking light on the ſubject. He denies, p. 374, that baſalt is volcanic, and with his uſual confuſion he ſubjoins an account of tufa and ſtalactite.

His

His new title of tracts on several subjects, &c. is abandoned at p. 410, where he assumes another title, that of *singular observations and improvements*, many of which are in fact idle theories, and none of them having any connection with mineralogy; it is unnecessary to give any detail of these heterogeneous digressions.

GLOSSARY

OF THE TERMS USED BY

MINERS IN DERBYSHIRE.

A

ADIT. A level. See Sough.

Arched. Arched. The roads in a mine when built with ſtone, are generally arched.

B

Bar-Maſter. An officer who ſuperintends the miners.

Barmote. A hall or court in which trials relative to miners are held.

Baſſet. When a ſubſtance as coal appears at the ſurface, it is ſaid to baſſet.

Belland.

Belland. Dufty lead ore.

Bit. A piece of fteel placed on the end of the borer.

Bind. A name given by miners to any indurated argillaceous fubftance.

Binghole. A hole through which the ore is thrown.

Bingplace. Where is laid the ore ready for fmelting and meafuring.

Blaft. When a hole is made with a borer of fufficient depth, it is filled with gunpowder to force off the rock, and the procefs is called blafting.

Borer. A round piece of iron three quarters of an inch in diameter, and two feet long, fteeled at one end with a fhort flat edge.

Bowfe. Lead ore, as cut from the vein.

Bucker. A piece of iron about fix pound weight, with a wooden handle, ufed for breaking the bowfe.

Buddle. A frame made of wood and filled with water.

Budling. Wafhing inferior lead ore, to free it from extraneous matter.

Bunding. Wood placed to throw the refufed cuttings on, or deads.

C

Catdirt. A fubftance fometimes called toadftone, being fometimes a variolite, at others a kind of limeftone.

Cart.

Cart. A machine uſed to draw ore, &c. out of the mine.

Chair. Uſed in drawing up ore or coal.

Cleanſer. A wire uſed after boring, to clear the hole.

Clevis. An iron at the end of the engine rope, on which the bucket is hung.

Coeſteads. A ſmall building.

Cope. To agree to get ore at a ſixed ſum per diſh, or meaſure.

Coper. One who agrees to take or make a bargain to get ore.

Corf. A kind of ſledge uſed to carry ore from the miners at work, to the drawing ſhaft foot.

Croſs veins. Veins that croſs each other.

Croſs Cuts. Are driven diametrically acroſs the range of the vein.

Croſſes and Holes. When a perſon diſcovers a vein, and has no means to poſſeſs it for want of ſtowces, he marks the ground with croſſes and holes, by which means he poſſeſſes it until he can procure ſtowces.

D

Deads. Cuttings of ſtone of no uſe.

Dial. A compaſs.

Dialing. The taking the different bearings of the various ways, gates, &c. in a mine, in order to ſink a ſhaſt from the ſurface on any particular ſpot with exactneſs.

Difh. A meafure containing 15 pints Winchefter meafure.

Due. The fame as lot.

Door. A crofs cut for a door is fometimes ufed to open and fhut, to increafe the circulation of air.

Drift. The place the miner excavates to make a road.

Driving. Cutting and blafting horizontally.

Ditch. A drain made at the furface to carry water off.

F

Fang. A cafe made of wood, &c. to convey wind into the mine.

Faufted. Refufe lead ore, to be dreffed finer.

Fault. A fiffure which breaks the ftratum.

Feigh. The refufe wafhed from the lead ore.

Flat. Flat work, when a vein, &c. is horizontal.

Forks. Pieces of wood, ufed to keep the fide up in foft places.

Foundermere. The firft 32 yards of ground worked.

Founderfhaft. The firft fhaft that is funk.

Forefield. The face, or vein worked,

Freeing. Entering a mine or vein in the bar-mafters book.

Fuzze. Straws, or hollow briars, reeds, &c. filled with powder.

Fuzze-borer. An iron made red hot to bore a fuzze to hold powder.

G.

G

Gallery. A drift or level.

Gate. The ſame.

Gears. Uſed to the cart, a kind of harneſs for the men that draw ore out.

Grove. A mine.

Gingonin. Walling up a ſhaft inſtead of timbering, to keep the looſe earth from falling.

H

Hade. The underlaying or inclination of the vein.

Hadings. When ſome parts of the vein incline, and others are perpendicular.

Hangbench. Part of the ſtowces.

Hangingſide. The ſide a vein hangs to.

Horſehead. A large opening made of wood, to turn and put on to a fang or trunk, to convey wind from day-light.

J

Jig Pin. A pin uſed to ſtop the machine in drawing when neceſſary.

Jumper.

Jumper. Borer, an iron inſtrument from a foot to three feet long, one end of which is ſteeled and worked to an edge.

K

Kevel. A ſparry ſubſtance found in the vein, compoſed of calcareous ſpar, fluor, and barytes.

Kibble. A bucket uſed for drawing up ore out of the mine.

Kit. A wood veſſel of any ſize.

Knits. Small particles of lead ore.

Knockings. Lead ore, with ſparry matter as cut from the vein.

Knockſtone. A ſtone uſed to break the ore on, but ſometimes it is a piece of caſt iron.

L

Leap. The vein is ſaid to leap when a ſubſtance interſects it, and it is found again, a few feet from the perpendicular.

Leadings. Small ſparry veins in the rock.

Level. An Adit, gallery, or ſough.

Limp. An iron plate uſed to ſtrike the refuſe from the ſieve in waſhing lead ore.

Loch. A cavity in a vein.

Lot. A certain proportion taken for the lord of the manor.

M

Maul. A large hammer.

Mear. Thirty-two yards of ground on the vein.

Metal. A word ſometimes uſed to expreſs an indurated clay above ſalt and coal.

N

Noger, or Jumper. See jumper or borer.

O

Old man. Places worked centuries ago, or in former ages.

O'erlayer. A piece of wood uſed to place the ſieve on, after waſhing the ore in a vat.

Opens. Large caverns.

Opencaſt. When a vein is worked open from the *day*.

Ore. The mineral as produced in a mine.

P

Pee. A piece of lead ore.

Pipe. A vein running unlike a rake, having a rock roof and ſole.

Plumb. A line and lead to meaſure depth.

Poſſeſſion,

Poſſeſſion. When ſtowces or wooden frames are placed on a vein, it is ſaid to be in poſſeſſion.

Pricker. A thin piece of iron uſed to make a hole for the fuzze to fire a blaſt.

Primgap. A variable diſtance, between two poſſeſſions.

Poling A plank or piece of wood, to prevent earth or ſtone from falling.

R

Rake. A perpendicular vein.

Ratchell. Looſe ſtones.

Rider. A rocky ſubſtance that divides the vein.

Riſing. A man working above his head in the roof, is ſaid to be riſing.

Roof. The part above the miners head.

Rubble. Same as ratchell.

Run. When the earth falls, and fills up ſhafts or works, it is ſaid to run.

S

Scaffold. In a mine, a platform, made where ſome miners work above the heads of others.

Scrin. A ſmall vein.

Shot. Blaſting.

Sled. A ſledge to draw ore without wheels.

Scraper. A ſmall iron uſed to ſcrape the ore, a kind of rake.

Shakes. Fiſſures in the earth.

Shift. The time a miner works.

Shaft. A perpendicular hole cut to get up the ore.

Sinking. Working deeper.

Smelting. Reducing the ore to metal.

Smitham. Small lead ore, duſt.

Smut. Decompoſed dark earthy ſhiſtus.

Sole. The bottom of the mine.

Sole Tree. A piece of wood belonging to ſtowces, to draw ore up, from the mine.

Sough. An adit or level.

Spindle. A part of the drawing ſtowces.

Stickings. Narrow veins of ore.

Stimmer. A piece of iron uſed to ram the powder with, when a blaſt is intended.

Stemples. Wood placed to go up and down the mine inſtead of ſteps.

Strings. Small veins of ore.

Stope. A piece of mineral ground to be worked.

Stopeing. Cutting mineral ground with a pick.

Stowces. Drawing ſtowce, a ſmall windlaſs.

Stowces. Pieces of wood of particular forms and conſtructions placed together, by which the poſſeſſion of mines is marked——a pair of ſtowces poſſeſs a mear of ground.

Sump. A ſhaft or perpendicular hole under ground.

Swallows. Caverns or openings where the water loſes itſelf.

T

Trogues. Wooden drains like troughs.

Troubles. Faults or interruptions in the ſtratum,

Trunks. Wooden ſpouts to convey wind or water.

Turntree. A part of the drawing ſtowces or windlaſs.

U

Underlay. When a vein hades or inclines from a perpendicular line, it is ſaid to underlay.

V

Vein. Any ſubſtance different from the rock, a rake vein is perpendicular, a pipe nearly horizontal.

Vat. A wooden tub uſed to waſh ore and mineral ſubſtances.

W

Walling. When the roads in the mine, are made with ſtone, it is called walling. The ſides of the mine or gangart, is frequently called the wall.

Waſh-hole. Where the refuſe is thrown.

Water-holes. Places where the water ſtands.

Weighboard. Clay interfecting the vein.

Wedge. An iron tool to get ore, fplit rocks, &c.

Wim. An engine or machine to draw ore worked by horfes.

Wind-holes. Shafts or fumps, funk to convey wind or air.

Windlafs. A well-known machine ufed to draw up ore. See Stowces, by which name it is commonly called.

Windlefs. A place in a mine where the air is bad or fhort, it is then faid to be windlefs.

Y

Yokings. Pieces of wood afcertaining poffeffion. Stowces.

THE END.

LONDON: PRINTED BY W. PHILLIPS,
GEORGE YARD, LOMBARD STREET.

Printed in the United States
By Bookmasters